Calculators

Calculators

Readings from the *Arithmetic Teacher* and the *Mathematics Teacher*

edited by

BRUCE C. BURT
West Chester Area School District
West Chester, Pennsylvania

National Council of Teachers of Mathematics

Library of Congress Cataloging in Publication Data:

Main entry under title:

Calculators.

Bibliography: p.
1. Calculating-machines—Addresses, essays, lectures.
2. Mathematics—Study and teaching—Addresses, essays, lectures. I. Burt, Bruce C. II. Arithmetic teacher. III. Mathematics teacher.
QA16.C34 510'.28'5 79-17365
ISBN 0-87353-144-2

Printed in the United States of America

Contents

2. Calculators and the Curriculum 53

3. Calculator Activities 97

4. Research

Introduction

The hand calculator made a significant contribution to mathematics education in the seventies, and it will probably have an even greater impact on the teaching of mathematics in the future. Many articles and books have been written about calculators during the past few years, and the National Council of Teachers of Mathematics has been one of the leading publishers in this area.

Some authors have suggested that calculators should not be used until a student reaches high school; others have recommended them for preschool students. However, most mathematics educators, if not all, will agree that the calculator is an effective teaching tool at some point in the mathematics learning continuum.

The articles included here have been selected from the *Arithmetic Teacher*, the *Mathematics Teacher*, and the 1977 NCTM Yearbook, *Organizing for Mathematics Instruction*. Most were published no earlier than 1976. In November 1976 the *Arithmetic Teacher* published a special issue on minicalculators; most of the articles from that issue are included.

This collection should provide a meaningful and useful resource for anyone who is presently using calculators in the classroom or considering their use as a teaching aid. Calculators are here to stay and it is our job as mathematics educators to provide for their most effective use in teaching mathematics.

1

Calculators: Policies, Uses, and Selection

This section contains several viewpoints about the uses of hand calculators as well as suggested policies and some considerations in selecting calculators. The lead article by Immerzeel illustrates classroom uses of the calculator—even when only one is available. Shumway discusses the pros and cons of using hand calculators in school mathematics, and Machlowitz offers some thoughts on their use in the classroom.

Munson recommends that every district develop a system-wide policy on calculators so as not to confuse children. The earliest article in this book (Hawthorne, December 1973) suggests that the hand calculator will permit teachers to introduce such concepts as decimals and negative numbers earlier in the school mathematics program.

Once a policy is established and the uses of a calculator are considered, it becomes necessary to make an appropriate selection. Caravella provides a checklist for selecting the right calculator and Rogers offers some useful suggestions to educators who are considering large-scale purchases. A reprint from the Association of Teachers of Mathematics, edited by Leake, offers a test of each of eight specifications for electronic calculators.

Aidala and Rosenfeld suggest some ideas for using calculators in the middle school classroom, and Drake provides some specific ideas for using them in the elementary classroom. The results of a survey by Palmer give us some indication of what teachers think about minicalculators in the classroom. Usiskin suggests that using the calculator as a crutch is not necessarily a bad thing.

The final article in this chapter is from the 1977 NCTM Yearbook. Bell, Esty, Payne, and Suydam provide an overview of some recent reports on calculators, and they summarize the recommendations of these reports.

The hand-held calculator

by George Immerzeel

One out of ten people in the United States has a minicalculator. When you walk through the grocery store you see shoppers, with calculators in hand, making decisions about the best buys. Your banker may use a minicalculator to answer some of your questions. At the calculator counter in your favorite department store, you see clusters of customers examining calculators to buy for themselves or their friends. There is no question that more and more people are using a calculator. It is time for each teacher of mathematics to have a minicalculator and it is time that we, as teachers of mathematics, find ways to use these amazing little devices to teach mathematics.

Even one calculator in a classroom can be helpful. Let me share a few of the ways I have found to use one.

First, put your calculator in the interest center of your classroom. Take a few minutes once in a while to let students tell what they have learned from using the calculator. Put a problem on a card in the mathematics corner and post the answer on your bulletin board. For example: Make the calculator count to 1000 by ones. How long does it take? How long would it take to count to a million?

Second, give the calculator to one student to use during your class discussion. Use this student as a resource to help calculate answers when the need arises in the discussion. For example, the calculator can be used to verify the position of the decimal point in multiplication or to change a fraction to a decimal.

Third, use a calculator to extend an activity into the real world. Suppose you are studying area—you have been finding the area of things described in the book. Set up a lesson where teams of students make the necessary measurements to find the area of the classroom door or chalkboard or anything that has a convenient rectangular shape. Give one student the calculator and have each team ask that student for the answer to the computation necessary to find the area. For example, after the team finds that the dimensions of the door are 81 cm by 200 cm, the calculator student uses the calculator to find the product of 81 × 200.

Fourth, let students use the calculator in some games. Have a race. Put one student with the calculator. Give a name for a number, like 9 × 7 or 13 × 10. The calculator student computes the answer on the calculator and then reads it out loud. The other students try to write down the answer before the answer is read. Another useful game develops skill in estimation. Again, let one student have the calculator. Give an example like 25 × 62 and have the calculator student compute the answer while the others estimate an answer. Commend students whose estimates are within your acceptable limit.

Calculators can be used along with practice in mental arithmetic. You can give a problem orally: 6 + 2 + 5 − 3 + 6 − 10. The calculator is used to verify the students' mental computation.

There are a number of ways that one calculator can make the teaching task easier and more interesting. The calculator becomes a portable answer key that has several advantages over the usual answer key. If you use your calculator when you check the students' papers, it can help you identify the particular type of mistake the student is making. For example, you can readily check partial products in the multiplication algorithm, or recognize equivalent fractions in checking fractional answers. The calculator is also a real help when you are making worksheets or designing other activities for the students.

I suggest one caution when using the calculator. The temptation is to have students use the calculator to check complex paper-and-pencil computation. Put yourself in the position of the student who has just spent five minutes dividing a three-digit divisor into a five-digit dividend and then checks the answer in a few seconds on the calculator. This student is convinced that paper-and-pencil computation is foolish and that the school is unreal.

Some people have avoided using calculators in the classroom because they are afraid they will be "ripped off." Reasonable caution is necessary when using any teaching aid, but I have used many calculators in the classroom—over a period of several years and at various levels—and have not lost a calculator.

Some teachers feel that the student who uses the calculator will not develop the necessary pencil-and-paper skills. So far, I have not observed the development of any dependency on the calculator. I have found that students will undertake much more complex problems and that students can solve verbal problems using the calculator at about three times the rate the problems could be solved with pencil and paper alone.

These are but a few of the ways I have used a calculator. I suggest you try one in your classroom. If you do, I am sure you will also think of many more ways in which the calculator can be effectively used. A calculator will add a new interest for both the students and yourself. Of course, if you have more than one calculator you can find still other ways to use them.

Although now is not time for every teacher to have a calculator for each student in every class, it is time, from my point of view, to explore ways to do a better job of teaching so we will be ready for the time when every student has a hand-held calculator.

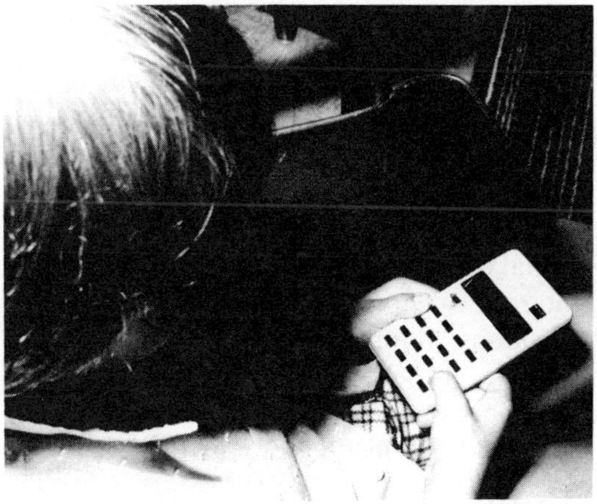

Hand calculators: where do you stand?

RICHARD J. SHUMWAY

A professor of science and mathematics education at the Ohio State University in Columbus, Dick Shumway is involved with the research, graduate, and undergraduate mathematics education programs.

It has often been proposed that the *Mathematics Teacher* and the *Arithmetic Teacher* publish careful arguments for and against a particular proposal to stimulate thought and air honest differences regarding issues relevant to the teaching of mathematics. Such *pro* and *con* discussions have been difficult to organize. At the risk of being accused of being on the wrong side of an issue by everyone, I would like to raise what appear to be the arguments proposed *for* and *against* the use of hand calculators in school mathematics.

ARGUMENTS FOR USING CALCULATORS

Proposal: Hand-held calculators as sophisticated as the scientific calculator (with such functions as arc, sin, cos, tan, hyp, lnx, e^x, log, x^2, $\sqrt{x}$, $1/x$, $x!$, π, y^x), should be made readily available to all children for school work, K through 12.

In support of such a position, variations of the following arguments may be cited.

1. *Paper-and-pencil algorithms for the basic operations will no longer be needed.* Algorithms are designed to carry out repetitive calculations efficiently, accurately, and without thinking. Clearly, the hand-held calculator is the best calculational algorithm available today. Paper-and-pencil algorithms might be taught for historical, cultural, or pedagogical purposes; however, few children (or adults) will choose paper-and-pencil algorithms when calculators are available.

2. *Scientific calculators will not be expensive.* The price of scientific calculators began only a few years ago at $400; currently, they are available for as little as $50. There is no reason to believe that they will not soon be available for less than $20 (which is the cost of two tanks of gas for a car). Costs will not significantly deter the widespread use of hand-held calculators.

3. *Extensive drill and practice exercises will be unnecessary.* Most children will probably learn the basic addition, subtraction, multiplication, and division facts in order to make estimations and to save time.

This paper was supported, in part by a grant from the National Science Foundation to Marilyn N. Suydam, The Ohio State University. A preliminary draft of the paper was prepared for the Ohio Secondary Mathematics Conference, September 1975.

4. *Decimals and scientific notation will be introduced early in the first grade.* Children will work with numbers such as .0285714285 and 1.893456 08. The first number is that part of a cake each of 35 children would get if the cake is divided into 35 equal parts; the second is approximately the number of seconds a six-year-old child has lived. Children can and will work comfortably with such numbers. Calculators will facilitate early continuous experiences with a whole new class of numbers.

5. *Mathematical exercises will be more realistic.* Exercises will no longer have to be chosen so that there are interger solutions. "Grubby numbers" and tedious calculations will be done with ease.

6. *Calculators are fun.* The motivational aspects of the hand-held calculator are exciting. Children create their own interesting problems. Low achievers generate new enthusiasm for mathematics because they finally have no fear of being unable to perform the necessary calculations. Children are eager to do mathematics when hand-held calculators are available.

7. *Addition and multiplication algorithms for fractions can be delayed until algebra.*

8. *Hand-held calculators facilitate number sense.* Because of their simplicity and speed, hand-held calculators will allow children to explore products, sums, powers, logarithms, trigonometric functions, and so on, with numbers of all sizes with a frequency never before possible. Intuitive number sense will be much facilitated by such extensive, continuous, and early experience with numbers and their properties.

9. *Hand-held calculators make calculations easy and practical for all children.* It must be remembered that decimal notation, Arabic numerals, zero, paper-and-pencil algorithms, and so on, were not introduced to teach mathematics, but to make calculations easier. The hand-held calculator was invented for the same reason.

10. *Hand-held calculators stimulate interest in and facilitate the teaching of mathematical concepts.* Homomorphic properties of functions, properties of logarithmic and exponential functions, characteristics of rational exponents, compounding continuous interest, combinatorics, trigonometric functions, limits, number theory, and so on, can all be learned in more interesting ways because of the calculational power that the hand-held calculator provides.

11. *Hand-held calculators can be used to facilitate problem solving.* Open exploration and new problems can be offered to children because of the facilitating calculational power that the hand-held calculator provides. For example, learning to predict for what interger values of x will $1/x$ fill the calculator display screen teaches a great deal about our base-ten numeration system and relative primes.

12. *Hand-held calculators provide experience with the only practical algorithm that is used in society today.* No business or profession carries out extensive calculations without the use of a calculator. Most family financial calculations will soon be done by calculator.

13. *Hand-held calculators will place the emphasis on when and what operation to use rather than on how to perform the paper-and-pencil algorithm correctly.*

14. *There will be more interest in estimation.* Since calculator errors tend to be dramatic rather than minute, estimating "ballpark" answers will be useful in avoiding errors.

15. *People's uses of mathematics will increase astronomically.* A simple example illustrates this. Suppose it takes $10 000 per year for a particular couple to retire today. Assuming an annual inflation rate of 5 percent, how much per year would be required 20 years from today? The sequence 1.05, [y^x] 20, [×], 10 000, [=] gives the answer of $26 533 in 10 seconds. Tailor-made family financial planning would be much improved by such calculational power.

16. *More time will be available to teach mathematics in depth.* Since calculators increase the speed and accuracy with which children can do calculations, much more time will be available to learn the concepts and principles of mathematics.

17. *New topics in mathematics can be introduced into the curriculum.* The calculational power of the calculator allows the consideration of new topics while the reduction in paper-and-pencil algorithms produces time for more topics.

ARGUMENTS AGAINST USING CALCULATORS

Proposal: Hand-held calculators should be banned from classroom use for mathematics.

In support of such a position variations of the following arguments may be cited.

1. *Hand-held calculators would destroy all motivation for learning the basic facts.* Hand-held calculators do not remove the need to know basic facts such as 9×7. To raise children to run to their calculators for every simple calculation would be folly. Such dependence on hand-held calculators would be most unfortunate.

2. *The use of hand-held calculators would destroy the basic, mainstream mathematics of the elementary school curriculum.* Society's major objective for elementary school mathematics is that children learn the basic facts and be able to perform the paper-and-pencil algorithms for addition, subtraction, multiplication, and division. If hand-held calculators are allowed in schools, children will no longer see any need for basic calculational skills. Even banning calculators on certain days or only using them for checking would seem unfair and illogical to children. Hand-held calculators must not be used for any teaching of mathematics.

3. *The cost of hand-held calculators makes their use prohibitive.* Schools simply cannot afford to provide calculators for children. The cost of hand-held calculators is too high and their attractiveness makes them disappear all too frequently.

4. *Hand-held calculators are particularly inappropriate for slow learners.* What possible motivation would such children generate for learning an algorithm that they know they can do on a hand-held calculator much more quickly and accurately? Hand-held calculators would insure that poorly motivated students would not learn the basic skills.

5. *The child's notion of the nature of mathematics would be changed by the use of hand-held calculators.* There is a real danger that if hand-held calculators are used, children will think that pushing buttons on a black box is mathematics.

6. *The use of hand-held calculators would reduce children's ability to detect errors.* Many people believe that if a calculation has been done on a calculator it must be right. Not only is such faith unjustified, but discovering errors of key-punching a calculator is almost impossible since there is no record of what was done.

7. *Paper-and-pencil algorithms are still necessary, basic skills.* Hand-held calculators can never be everywhere. Children must still be able to calculate on their own. The availability of calculators in schools would remove children's need for practicing the basic skills. Homework done at home would no longer ensure facility with the basic skills, since the home is likely to have a calculator. Schools must ban the use of calculators to ensure facility with the basic skills of arithmetic.

8. *Batteries lose their charge and wear out.* Dependency on batteries for computational arithmetic would be foolish.

9. *The use of hand-held calculators would discourage mathematical thinking.* If children can do any mathematical calculation by pressing a few buttoms, problem solving will be done by guessing, not mathematical thinking—try this, try that, keep doing

things with the numbers until the answer looks right. Nonthinking guessing will become rampant if hand-held calculators are available in schools.

10. *Parents are opposed to the use of hand-held calculators in the schools.* The schools have failed miserably in the teaching of basic skills as it is. The introduction of calculators would be, in effect, not teaching mathematics at all. Schools would be exhibiting extreme political ineptness to introduce hand-held calculators.

SUMMARY

The proponent's argument is essentially the following:

> The hand-held calculator is the tool used in society today for calculations. Schools are "burying their heads in the sand" if hand-held calculators are not recognized and used as the calculational tool that they are.

The opponents' argument is essentially the following:

> The principal objectives of mathematics instruction (at least in K–9) are that children learn the basic facts and the paper-and-pencil algorithms. Such learning will not occur if hand-held calculators are made available in the schools.

It would seem that a rational approach to the resolution of the problem would involve two questions:

1. What are the current and future societal needs for the basic facts and the paper-and-pencil algorithms?

(*a*) If there are no needs for such skills, drop the emphasis on them and introduce the widespread use of hand-held calculators.

(*b*) If there are needs for such skills, move to question 2.

2. Can the hand-held calculator be used in the classroom and still build students' needed skills (as identified in 1*b*)?

Such a procedure would seem to satisfy the concerns of the opponents of the use of hand-held calculators. The proponents of the use of hand-held calculators would probably claim that such an oversimplification of the benefits of the use of hand-held calculators is ignoring a potentially powerful educational device.

The Board of Directors of the National Council of Teachers of Mathematics has adopted the following position:

> With the decrease in cost of the minicalculator, its accessibility to students at all levels is increasing rapidly. Mathematics teachers should recognize the potential contribution of this calculator as a valuable instructional aid. In the classroom, the minicalculator should be used in imaginative ways to reinforce learning and to motivate the learner as he becomes proficient in mathematics. (NCTM, *Newsletter,* December, 1974)

In "Minicalculators in Schools," the NCTM Instructional Affairs Committee listed and illustrated by examples nine justifications for using calculators in schools. (*Arithmetic Teacher,* January 1976, pp. 72–74.)

The arguments have been outlined. Where do *you* stand?

ELECTRONIC CALCULATORS—FRIEND OR FOE OF INSTRUCTION?

Bombarded with questions on pocket calculators? Here are some thoughts and references on their purchase, and on their use in the classroom.

By ELEANORE MACHLOWITZ

Kensington High School
Philadelphia, Pennsylvania

THE most lasting revolution in mathematics education may prove to be the one caused by the proliferation of low-cost, compact electronic calculators. Anticipating the day when they share pocket and handbag space with comb and credit cards, some authorities see a lessening need for arithmetic instruction. This is not so, because correct answers depend on insights into the problem in order to choose the proper operations with the necessary numbers in the required sequence. So far, the issue of calculators and education is usually in the news as colleges and high schools make decisions about permitting their use as a substitute for arithmetic calculations in science and mathematics.

Recommended mathematics classroom uses have often been confined to eliminating lengthy computations in such areas as trigonometry, statistics, probability, and business problems. Calculators have also been used for pupil self-checks of algorithmic computation.

More significantly, the calculator can also present dramatic, attractive, and speedy opportunities for discovery, demonstration, and reinforcement in the general mathematics classroom with even the lowest-ability student. In fact, the slower student, usually resistant to deductive logic and abstractions, yet too sophisticated for arrays of ducks and portions of pies, can profit most.

The public school system of Berkeley, California has purchased pocket calculators for underachieving junior high school students, but according to John Kelley (Schwabach 1974), "The calculators will only be used for students too slow to handle math problems any other way." The Math Action Program in Philadelphia includes in its multifaceted program two calculators as teaching aids for each senior high school with pupils who need extra motivation. *Time* (6 January 1975) mentions many other school systems using classroom calculators, but in most instances they are used for computation rather than as a teaching device.

The classroom organization for using calculators as instructional aids can vary—a math lab with task cards, demonstrations to groups or the entire class, or, if a number are available, with small teams working together on specific assignments.

In most cases, the teaching strategy is an experimental approach. The student or demonstrator is directed to solve the same type of arithmetic problem illustrating the concept with a succession of varying inputs. The results are recorded on student worksheets, the blackboard, or a transparency. With the focus away from time-consuming computations, a pattern leading to a rule or shortcut quickly emerges. The students then formulate the newly discovered property or method and record this in their notebooks, worksheets or permanent wall charts.

It should be remembered that using a calculator for introductory and motivational purposes does not eliminate the need for repetitive drill to ensure retention of the skill or concept. Therefore, the next activity of the lesson plan is to provide paper-and-pencil practice, with answer checking by calculator if desired. More practice for homework and subsequent review provides additional reinforcement.

The following are some of the questions that would be answered by this inductive approach after a series of calculations:

1. Multiply a sequence of whole numbers by 10. Is there a short cut? Do the same with 100, 1 000.

2. Multiply a sequence of numbers containing decimals by 10, then by 100, then by 1 000. Compare the results.

3. Divide a sequence of numbers by 10, 100, 1 000.

4. Take several numbers and multiply each by .4, .40, 0.4. Are the answers different?

5. What happens to the decimal point in multiplication? Make several calculations to check your hunch.

6. Do the same with the division of decimals.

7. Multiplying by 0.5 turns out to be the same as multiplying by what fraction? And also as dividing by what number? Why?

8. Follow the same procedure for multiplying by 0.25. Now we have another way to derive a table of fraction-decimal equivalents.

9. A fraction then is another way of writing what operation?

10. Which of a group of fractions has the greatest value? Check your guess.

11. How can you classify the decimal equivalents of fractions with prime denominators?

12. Starting with 50%, 25%, and 10%, add a third column to our table of equivalents. (Avoid using a % key in order to emphasize the meaning of percent as a decimal equivalent or a third way of writing a fraction).

13. Compile a classroom chart of powers and (by repeated approximations and corrections) roots. (Here again, avoid using a radical key and a power constant, at first, to reinforce understanding of the terms.)

Other possible tasks and demonstrations include:

1. "Discover" most of the needed axioms and properties.

2. Investigate rules of divisibility.

3. Generate composite numbers.

4. Work with repeating decimals.

5. Use as a mechanical judge in mental arithmetic games.

In an article in the *New York Times* (5 January 1975), Edward B. Fiske reported that J. Fred Weaver of the University of Wisconsin found calculators useful as a teaching tool at an even lower level: "It allows a child to confirm that 29 plus 29 plus 29 is the same as three times 29."

For classroom use, table-top rather than pocket-size models are recommended for ease of operation, visibility for more students, and for prevention of theft. Some schools handle the security problem by mounting calculators on large boards, by securing them to desks with metal brackets, or by buying those with a security lock and chain package. For the vast majority of instructional purposes, a floating rather than a fixed decimal point is necessary, and eight digits on display should suffice. For the instructional methods outlined above, no keys other than the basic operations are required. An AC machine saves battery expense, but using several in a classroom will necessitate a maze of extension cords and octopus outlets. Some brands are both AC and battery-operated. Rechargeable batteries are of questionable value, since they are extremely expensive and not easily obtained at the present time.

The features cited, and more, can be found in models selling for under $25. Tape printout models would have advantages but are more costly. It is predicted that small pocket models will soon be selling for under $10. Preference should be given to a dealer with a one-year, over-the-counter replacement guarantee. Ratings and more general information can be found in an article in the June 1973 issue of *Consumer Reports*. More recent evaluations, along with a discussion of "Logic" and a "What the Keys Do" section are in the September 1975 *Consumer Reports.* Some practical buying and using tips are given in the October 1974 issue of *Money* magazine. John Free, in an especially informative article in the February 1975 issue of *Popular Science,* gives a tabulation of 34 features and functions for 37 well-known makes and models. The fast-paced changes in models and prices make evaluations outdated by the time of publication. Purchase decisions might best be based on visits to reliable discount and department stores, as well as on writing to the larger mail-order houses.

In many school systems, teachers interested in working with calculators will find that there is still a question whether they can be purchased under regulations governing expenditures. Sources other than budgeted requisitions that might be tried are student fund raising, PTA and alumni gifts, local foundation grants, and federal funding, if available.

Classroom use of calculators is so new that determination of long-range effects and the question of lasting increases of pupil achievement await extensive research. One early study (Cech 1972) with calculators used only to verify answers to whole number operations found no improvement in computational skills or attitudes toward mathematics. But, as he says, "It was not used to build insight into the understanding of mathematical principles." However, the immediate surge of interest and participation which is evident to any teacher who has brought an electronic calculator to class gives an added value to this new teaching tool which can help a teacher to introduce concepts.

REFERENCES

"Calculators." *Consumer Reports* 40 (September 1975): 533–41.

"Calculators in the Classroom." *Time* 105 (6 January 1975): 88.

Cech, Joseph P. "The Effect of the Use of Desk Calculators on Attitude and Achievement with Low-Achieving Ninth Graders." *Mathematics Teacher* 65 (February 1972): 183–86.

Comarow, Avery. "Practical New Uses for Pocket-Money Calculators." *Money* 3 (October 1974): 88–91.

"Electronic Mini-Calculators." *Consumer Reports* 38 (June 1973): 372–77.

Fiske, Edward B. "Educators Feel That Calculators Have Both Pluses and Minuses." *New York Times* (5 January 1975): p. 7.

Free, John R. "Now There's a Personal Calculator for Every Purse and Purpose." *Popular Science* 206 (February 1975): 78–81.

Schwabach, Robert. "Philadelphians Are Flipping Over Pocket Calculators." *Philadelphia Inquirer* (10 February 1974).

Your District Needs a Policy on Pocket Calculators!

By **Howard R. Munson**

A quick glance at professional journals dealing with teaching generally, and with mathematics teaching specifically, reminds us that miniature calculators have become a concern and an item of great interest. Articles dealing with the major question, "Should minicalculators be used in schools?" abound. Articles suggesting ways the miniature marvels may add to, or detract from, students' performance, understanding, and competence, are proliferating at a rate perhaps calculable only on a machine.

Within a single school building, perhaps even among teachers of a single grade within a school building, attitudes will vary as to whether children may use calculators in school or to do school assignments. Even among those who agree calculators should be used, there will likely be differences of opinion relative to who should use calculators and under what circumstances. What is needed is a school or district policy!

I urge that every teacher who is involved in mathematics instruction appeal to his or her administrators for an opportunity to develop a policy that will be operative system-wide. Do whatever is necessary to formulate a single policy that will govern the use of pocket-sized calculators and spell out who may employ them, when, and under what conditions.

Without a system-wide policy, there will be a hodgepodge of policies that will differ from teacher to teacher and school building to school building. Children will be confused and, if they should happen to transfer, may be placed at a disadvantage by not having had prior experience or training with a calculator. Parents will be confused and may well exercise pressures in one direction or another. A system-wide policy, well articulated, and well publicized, will alleviate many problems and will aid in the avoidance of others.

Formation of a system-wide policy will likely be best done by a committee representative of all grades and schools in the district. However, subcommittees may be necessary to consider the question adequately for elementary schools, and for junior and senior high schools.

Here are some questions to consider.

1. Should calculators be used for computation or merely to check on computations done by hand?

2. Should slow learners be given access to calculators in the belief that students will gain mathematical facility with their use?

3. At what grade level should calculators be introduced if they are to be used as a part of an instructional unit on calculator use?

4. Is there need for an instructional unit on calculator use?

5. Will the use of calculators make students mathematically lazy?

6. Should restrictions be placed on the use of calculators in subjects other than mathematics?

7. If calculators are used for daily work, may they be used in tests?

8. Should special allowances be made for the use of calculators in achievement test batteries?

9. Does the school have a responsibility to furnish calculators for student use?

10. Should calculators replace slide rules in advanced mathematics and science classes?

The foregoing list of questions is in no way exhaustive. It is meant only to open a few important areas for consideration.

Two additional recommendations relate to the establishment of system-wide policy. First, the school should establish a study committee to monitor and gather evidence relative to the manner in which the policy is functioning. Second, a date should be set for the evaluation and reconsideration of the policy governing the use of calculators. These recommendations, if followed, will assure that a viable, defensible policy will emerge. □

As professor of education at Winona State University in Winona, Minnesota, Howard Munson is responsible for science and mathematics methods courses for preservice elementary teachers. His current interest is in preparing teachers to teach remedial mathematics in the middle grades.

Hand-held calculators: help or hindrance?

FRANK S. HAWTHORNE

Describing himself as a "teacher, administrator, mathematician, and a lot of other things, in that order," Frank Hawthorne is Chief of the Bureau of Mathematics Education in the New York State Education Department.

For more than a decade mathematics educators have been on notice that the day is approaching when everyone will have ready access to sophisticated electronic computers. Recent advances in the development of small, inexpensive calculators provide impressive support for the prediction.

Calculators that are battery-powered and small enough to fit in a shirt pocket are now available for as little as $50.00. They are advertised as Christmas gifts for the person who has everything. Since competition between companies in the business of producing these little calculators is relentless, we may expect prices to drop even lower.

These reliable, sturdy little devices have the capacity to add, subtract, multiply, and divide. They can work with ten-digit numbers and operate with decimal fractions and negative numbers. What effect will the availability of these calculators have on the elementary school mathematics program? Few changes will be necessary. Currently elementary schools emphasize understanding of concepts and a meaningful approach to computational algorithms. Calculators can make only fringe contributions to these areas and the existence of the calculators does not detract from the significance and relevance of such goals. Indeed, it was partly the anticipation of inexpensive calculators that impelled curriculum designers to decide that drill on computation merely to develop speed in calculation was unwise. This point of view has strongly influenced most modern elementary school mathematics programs.

Today there are no jobs that require a clerk to do lengthy arithmetic calculations using pencil and paper. But it is still important to *understand* arithmetic. A hand-held calculator can save hours when you are doing your income tax statement, but it is useless to you if you don't understand, say, percent.

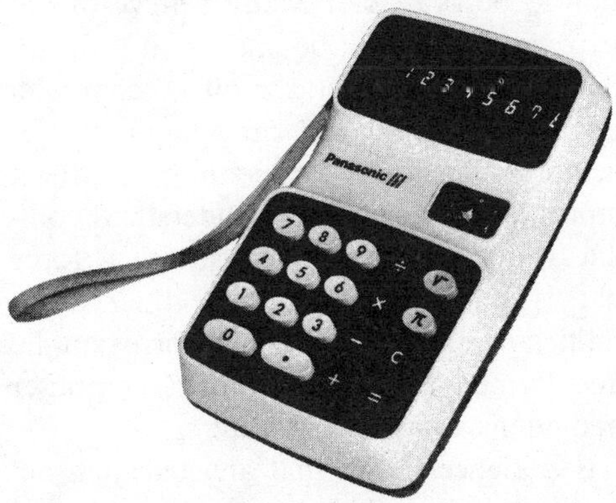

While producing no grand changes in programs, hand-held calculators offer important advantages to elementary school teachers. Obviously they make it possible for a youngster to check the accuracy of his answer thus providing immediate verification, which is an important motivational factor.

Far more significantly, hand-held calculators can eliminate tedious, unnecessary calculations that consume precious time and destroy interest. For example, a student doing a problem such as: 71,265 ÷ 29.6 is presumably proficient in subtraction and multiplication. He can use the calculator for the multiplication and subtraction involved in this problem thereby allowing himself to concentrate on the primary goal—learning to understand the division algorithm.

An important point to remember about hand-held calculators is that they permit students to get answers using operations they have not yet studied. Therefore, it is important for teachers to emphasize the importance of showing *how* an answer is obtained. It also becomes important to introduce decimal fractions and negative numbers early in the school mathematics program because users of calculators are going to meet them quickly. These facts have implications for mathematics supervisors as well as teachers.

Is the advent of the hand-held calculator a signal that the hand-held computer is soon to be available? Many believe this to be true. If so, it will be important to introduce flow charting and a bit of computer programming into the curriculum of elementary schools. It is becoming of critical importance for citizens to understand computers since computers constitute a force that is tending to dominate their lives. Politicians on our west coast, for example, are alarmed that election-night computer predictions sway many voters.

Experience shows that any teaching aid can be misused. Are there some potential educational disadvantages connected with the hand-held calculators? The answer is "probably yes" and teachers should consider in advance (with the aim of avoiding them) possible misuses of these devices.

If introduced too early, before a child has developed some "number sense" and familiarity with the basic operations of arithmetic, calculators could do great harm. As a "black box" that furnishes answers with no hint to the operator as to how these answers are obtained, they do not help students to gain the understanding of basic number concepts generally considered necessary.

We may expect to hear promoters of the use of these instruments in elementary schools draw parallels to the role of the abacus in oriental countries, where abacuses are used in all walks of life, including classrooms. It should be remembered, however, that the operations on an abacus can illustrate mathematical operations much better than an inscrutible electronic calculator.

"Hand-held Calculators: Tool or Toy?" was the title of an article in the *New York Times* on 20 August 1972. The hand-held calculator can be a very valuable tool, but only to an operator who understands the basic ideas, concepts, and meanings behind the instantaneously generated answers it provides.

In an attempt to determine what reasonable role these calculators may play in schools, the New York State Education Department has arranged for a trial of such calculators in two schools during the 1973–74 school year. Each sixth-grade student of one class in each school has been issued a calculator for use throughout the year. Control groups have been established, and pretesting and posttesting are planned. The results of this project will help to answer the question, What is the effect of having these calculators available to students?

Selecting a minicalculator

JOSEPH R. CARAVELLA

Previously a junior high mathematics instructor for both private and public schools in western New York, Joe Caravella is now Director of Professional Services for the NCTM.

As long as the number of minicalculators in use continues to increase and the quality of their hardware and software designs remains in a state of flux, a systematic means of differentiating between and evaluating the various machines will be needed. The checklists that follow are designed to help educators select an appropriate minicalculator for their individual or group situation.

The screened sections identify specific suggestions for the elementary classroom.

I. Consider both current and future computational needs

Use	*Category*
☐ Elementary school ☐ Junior high school ☐ General consumer	*Basic.* A minicalculator with the minimum capabilities listed in II.
☐ High school ☐ College	*Scientific.* A basic minicalculator with some special keys and additional features.
☐ Business	*Business.* A basic minicalculator with special algorithms and features most frequently used for financial and statistical applications.
☐ Post graduate ☐ Industry	*Advanced scientific.* A programmable calculator with several special keys and additional features.

II. Identify and evaluate the specific features of a minicalculator

Minimum capabilities and considerations

- ☐ Algebraic logic (See III.)
- ☐ Basic functions: [+], [−], [×], [÷]
- ☐ Floating decimal point—when **0.6666666** or **.66666666** is displayed on an eight-digit calculator after 2 [÷] 3 [=] is entered. If **0.6666667** or **.66666667** is displayed, the machine is rounding off in the last place.
- ☐ Overflow and/or error indicator—what happens when the following are entered: 999 [×] 999 [×] 999 [×] 999 [=], 7 [÷] 0 [=], and .1 [×] .02 [×] .003 [×] .0004 [=] ?
- ☐ Bright, easily readable eight-digit display with an acceptable viewing angle
- ☐ A negative sign that immediately precedes a negative number

- ☐ "Clear" and "clear entry" capabilities
- ☐ No keys with nonassociated second functions (for basic minicalculators only)
- ☐ Convenient keyboard format
- ☐ Reasonable key size for your fingers
- ☐ Springloaded, "click" keys
- ☐ Appropriate machine size and weight
- ☐ Battery and AC adapter options. Consider the number of operating hours per battery replacement or charging. Automatic power-down displays and delayed power-off features insure the maximization of battery life. Long-life replaceable batteries seem to be the most cost- and time-efficient. Charging batteries and contending with electrical cords can be tedious.

Special keys

	Key			Key	
☐	[K]	constant	☐	[x^2]	x squared
☐	[+/−]	change sign	☐	[%]	percent
☐	[$\sqrt{\ }$]	square root	☐	[$1/x$]	reciprocal
☐	[π]	pi	☐	[y^x]	y to the x power
☐	[( ] [ )]	parentheses	☐	[$\sqrt[x]{y}$]	xth root of y
☐	[$x!$]	x factorial	☐	[$x \leftrightarrow y$]	x exchange y
			☐		other

Additional options

- ☐ Memory
  - ○ Storage memory—stores the number displayed for later recall.
  - ○ Addressable memory—allows basic functions to be performed on and retained in the content of a memory register.
    - △ Memory indicator
    - △ "Memory clear" key
    - △ "Memory exchange" key
- ☐ Scientific notation
- ☐ Automatic constant—preferably the last function and number entered before depressing the [+].

  Examples:

8	[+]	2	[=]	10.	8	[×]	2	[=]	16.
13			[=]	15.	13			[=]	26.
5			[=]	7.	5			[=]	10.
			[=]	9.				[=]	20.
			[=]	11.				[=]	40.
8	[−]	2	[=]	6.	8	[÷]	2	[=]	4.
13			[=]	11.	13			[=]	6.5
5			[=]	3.	5			[=]	2.5
			[=]	1.				[=]	1.25
			[=]	−1.				[=]	.625

☐ Programming
☐ Print-out
☐ Other ____________

Additional functions

☐ Trigonometric
☐ Logarithmic
☐ Hyperbolic
☐ Other ____________

III. Decide on a logic system

Be alert for new logic systems and varying degrees of sophistication for those indicated in figure 1. Remember that identical keys on different machines may perform different functions.

Logic	*Keycharts for sample problems*	*Remarks*
☐ Algebraic ☐ Keyboard has an [=] key and separate [+] and [−] keys.	3 − 2 = ______ 3 [−] 2 [=] **1.** 8 × 2 − 7 = ______ 8 [×] 2 [−] 7 [=] **9.** 25 + 15 ÷ 5 = ______ 25 + (15 ÷ 5) = (15 ÷ 5) + 25 15 [÷] 5 [+] 25 [=] **28.** or 25 [+] 15 [÷] 5 [=] **28.**	Most appropriate because it allows data to be entered as you would write a problem. However, you *must* remember the order-of-operations for many machines. Some newer calculators are programmed to follow the standard algebraic rules and allow for varying levels of parentheses.
☐ RPN (Reverse Polish Notation) keyboard does not have an [=] key.	3 [ENTER] 2 − 1. 8 [ENTER] 2 [×] 7 [−] 9. 15 [ENTER] 5 [÷] 25 [+] **28.** or 25 [ENTER] 15 [ENTER] 5 [÷] [+] **28.**	Bracket-free notation. Popularly used for advanced scientific applications.
☐ Arithmetic Keyboard features [+=] and [−=] keys,	3 [+=] 2 [−=] 1. 8 [×] 2 [+=] 7 [−=] 9. 15 [÷] 5 [+=] 25 [+=] **28.**	Seldom found in newer minicalculators. Commonly employed by desktop business machines.

Fig. 1

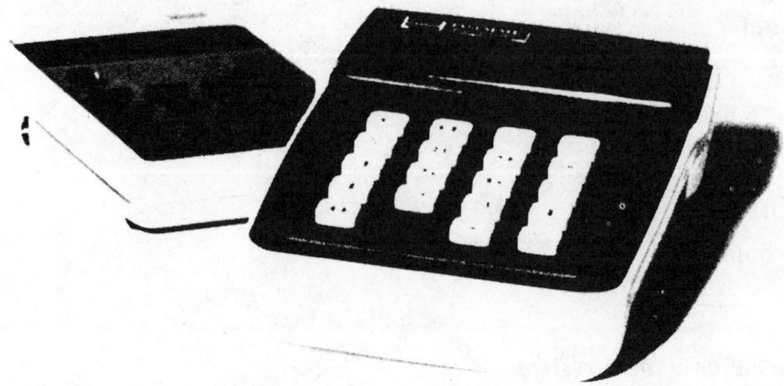

IV. Determine how much money is available to invest in minicalculators

With minicalculators priced below $10, they are available to most students.

Calculators suitable for elementary and intermediate classrooms range from $10 to $20.

Calculators used at the secondary level are usually less than $50, while the more advanced and specialized machines may cost several hundreds of dollars. Special minicalculators for use on an overhead projector and a talking minicalculator for handicapped students are available.

The number of calculators in a classroom is not of primary importance. Activities can be designed to fit any situation. You may begin with only one minicalculator in the classroom (probably your own); in the long run each student will have his or her own machine. Remember, you are living in a world of minicalculators.

A classroom set of the same calculator model would be the ideal situation at the elementary level.

V. Select a reliable manufacturer that provides at least a twelve-month warranty and a repair service for its products

Most malfunctions occur during the first thirty days and are usually due to a power source failure.

Now, using your nonwriting hand to enter data, enjoy your minicalculator.

The electronic calculator—another teaching aid?

JOY J. ROGERS

An associate professor of educational foundations at Loyola University of Chicago, Joy Rogers is interested in matters of instructional design and development.

Fifteen years ago schools seemed to find it very easy to get grants to install language laboratories. Seven years ago I attempted to do a paper (Rogers 1969) on what had become of all those language laboratories. The paper was quite a challenge because published references to language laboratories had dwindled to nearly none. An exciting, promising teaching aid had, in the span of only a decade, been relegated to the dustiest corners of the school along with the supply of Pressey punchboards, wire recorders, and instructional radio receivers. This paper is not just about electronic calculators—or even about language laboratories. The paper is intended to identify some issues that appear to be responsible for the all-too-rapid and unforeseen obsolescence of *any* teaching aid. It is hoped that the paper might offer some useful suggestions to the manufacturers of new teaching aids—in this case electronic calculators—in the hope that some of the mistakes of the past may not be infinitely repeated. Secondly, it is hoped that this paper might offer some useful suggestions to educators considering large-scale purchases of electronic calculators on how to avoid, ten years from now, being the butt of other administrators' jokes about their horde of these push-button white elephants.

The most critical point that must be made in such a paper as this is that the sad tale of the language laboratory need not be a forecast of the inevitable future of a teaching aid. The chalkboard, for example, is an example of a teaching aid that has been, and continues to be, extensively used in classrooms. Teachers may be somewhat apologetic about use of the chalkboard, explaining that the overhead projector is broken or that the children enjoy writing with chalk, but it is almost unthinkable that a school would be built *without* lots of chalkboards. Some chalkboard manufacturers have tried various modifications such as changing the color or painting permanent lines on them, but the method by which the chalkboard is used remains the same. It might be useful to consider some of the features that have made the chalkboard an enduring teaching aid amid so many bright-with-promise, but ephemeral, other devices. I propose a list of four features that seem to separate enduring teaching aids from the others.

Inexpensive and durable

First, the teaching aid must be inexpensive enough or durable enough for child use. The concept of inexpensive, here, is not intended as a relative term, but is absolute in the sense that the item must be either inexpensive or durable enough that most teachers would allow relatively unsupervised children to use it freely. The chalkboard, though expensive, is rather durable and is an excellent example of a freely used teaching aid in many classrooms. A potentially promising, but so far disappointing, teaching aid is the cassette recorder. It is almost ironic that cassette recorders have become inexpensive enough that many children own their own and learn a great deal from them by using them in play situations. School budgets, however, are apparently not as liberal as household budgets and cassette recorders are still expensive enough that the typical teacher watches over them meticulously. Though I have seen children out of school play freely with cassette recorders and children in school playing freely with chalkboards, I have never seen a child in school playing freely with a cassette recorder. Even the child who brings *his own* cassette recorder to school is likely to make his teacher upset if he experiments with it in any way or even appears as if he might drop it.

An obvious essential feature of any teaching aid is that it does actually teach—that is, that it responds to the learner's behavior in some consistent way that is satisfying or interesting to the learner. If one goes a step further in that line of reasoning, then it must follow that important things to be learned from any teaching aid are the *limits* of the medium. These can be learned in an interesting and surprisingly efficient way through exploration or "discovery" learning. For example, a child who has been making physical representations of three-dimensional objects while "playing" with paper will quickly discover one of the limits of the medium when he moves to the chalkboard. His dissatisfaction with his two-dimensional chalkboard drawings may, then, lead him to learn to draw in perspective. If he had only been allowed to use the chalkboard in a restrictive way—for example to write his spelling words—he might not have developed this other useful skill.

If a classroom is stocked with electronic calculators that are locked safely in a closet except during arithmetic class, at which time they are cautiously placed on tables to prevent any danger of their being dropped, then it is likely that the calculator may be as disappointing as the cassette recorder. To be a genuinely effective instructional tool, calculators must be inexpensive enough that they can be taken to the gym, the playground, the lunchroom, the band practice room, or any place where mathematical questions are of interest to learners.

Controllable by learner

The second feature of an enduring teaching aid, a logical outgrowth of the first, is that it can be controlled by the learner. The book is an excellent example of such a teaching aid. The design of the book had long given it the potential to be a useful teaching aid. It awaited only Gutenburg's presses to make it practical for books not to be chained to walls and read only by monks who lovingly caressed their pages. The reader controls many aspects of the operation of a book. He can read it at any rate and can refer to any part of it. The book can be read in a variety of different reading environments and body positions. The individual reader can stop or start it at will. In contrast, consider the 16-mm motion picture projector as a teaching aid. Most frequently the projector must be operated by either a teacher or a specially trained student assistant. It cannot be conveniently stopped for an individual learner because it is usually used to show films to large groups. It is difficult to make only a particular portion of a film accessible for reference. It is nearly impossible for a student to view a film anywhere other than in his classroom or at any time other than the assigned time. In consequence, it is easy to find schools relatively well supplied with books and chalkboards, but having only a few, rarely used motion picture projectors.

The point here is that an effective teaching tool needs to be at hand whenever there is a problem to be solved or an object of curiosity to be investigated. The electronic calculator has potential versatility comparable to that of the book. A learner can move it to and use it in about the same range of situations in which he might use a book. It is even amusing to speculate what the world would be like if people began to develop a "calculator habit" comparable to the "book habit" now a part of the lives of many. Image an ordinary individual thinking through some new mathematical concept with the aid of his pocket calculator—in his car while waiting for the traffic light to change, at breakfast in lieu of the morning paper, or even at night while soaking in the tub.

The size and shape of existing calculators make such uses quite possible. The only barrier is that most people who are now adults have grown up without having such mathematical freedom. Simply the calculations involved in solving practical problems are likely to be so forbidding that the adult may never have thought about "playing" with mathematics as a pastime. If, however, it is possible to produce a generation of children who can calculate any item of curiosity—from their own baseball averages to the amount of gas required to drive the family car a distance equivalent to around the world—without the computations being a burdensome chore, then perhaps they will grow to be adults who do enjoy mathematical thinking.

Tiny and cheap calculators ought not to present a problem for either educators or manufacturers. True, educators will have to begin to think of mathematics as something more than computation; but recent interest in modern mathematics, non-Euclidean geometries, and other nontraditional topics suggest that educators *are* eager to teach more than computation. Presumably, too, manufacturers should find a market created in this way virtually impossible to saturate. It seems likely that a child who became used to using a $3.98 shirt-pocket-sized calculator (for which he would not be spanked if he lost it) would soon want elaborate and specialized calculators to deal with his increasingly sophisticated questions.

Does something learner wants done

This leads to the third feature of an enduring teaching aid. It should solve problems (or do things) that the learner wants done. Good examples of such tools have been, for older learners, the slide rule and, for younger learners, Montessori-type toys. The slide rule has allowed the learner to estimate "how much" or "how far" about something he wants to know without waiting to do elaborate computations. Montessori toys allow a child to master certain relationships of size and number at a time when he is curious about those relationships. A nonexample of this kind of teaching aid might well be the linear teaching machine. The linear teaching machine is designed to present a series of questions and answers in a sequence predetermined by the author. Unlike the user of a book, the user of a teaching machine cannot look in the index and turn only to the particular information he needs to solve his particular problem.

The pocket calculator is, of course, the obvious successor to the slide rule for persons of ages or social circumstances to have been using slide rules even if pocket calculators had not been invented. Slide rules were complex enough that one had to know which scale to use for whatever function one wanted, at which end of the rule to read one's answer, and even where the decimal would have to go. These features made it prohibitive for use by ordinary nine-year-old who wondered how many packages of sunflower seeds he could buy if he had a million dollars. An electronic calculator could give the child such a capability.

An educational question might well be raised, *Should* children encounter mathematics in this way? One concern of cognitive developmental psychologists (Phillips 1969) has been that American educational practice pushes children too rapidly into mathematical processes for which they are not cognitively ready. Another observation might be made about the nature of the mathematical instruction children have received. They usually acquire adequate enough reading skills to pursue their own verbal interests by the end of the primary grades. Their mathematical instruction, however, remains abstract and not immediately related to their other problems and interests for many more years. It is interesting to speculate about what a few years of availability of electronic calculators will do to developmental psychologists' concepts about young children's cognitive structures.

Flexible usability

The final characteristic of an enduring teaching aid to be considered in this paper has to do with their manufacture. Built-in obsolescence and the necessity for specialized, compatible materials make teaching aids of limited use. The chalkboard again provides a good example. Attempts to manufacture chalkboards that require particular types of chalk have been eminently unproductive. There is virtually nothing else a manufacturer could do to make his equipment compatible only with the supplies he has manufactured. This was not the case with, for example, the teaching machine. Teaching machines are simply presentation devices and must have a supply of programs to be of use. The array of published programs that are set up in a way as to be compatible with only one manufacturer's machine is grim testimony as to how quickly healthy competition turns to simple greed. The obvious long-range impact of such a practice is that schools will avoid investing in any of the machines lest their presentation devices not be suitable for future programs.

The nature of the electronic calculator makes this kind of planned obsolescence as a teaching aid almost impossible. It is self-contained. If, as educators, we can avoid the all-too-tempting pitfall of trying to publish rigid "lesson plans" or "calculator exercises," then continued improvements or modifications in calculators will not make the old ones obsolete.

SUMMARY

If one is to try to predict the future, one ought to be wise enough to insert enough "if" clauses to cover most eventualities. Thus, I would like to summarize my position by suggesting that the electronic calculator has the potential, as assessed against these four criteria, to be a teaching aid of enduring value. If both educators and manufacturers will give continuing consideration to these points, we may be fortunate enough to see a generation of children as fluent with numbers as previous generations have been with words.

The price of failure to keep these considerations in mind is also quite evident. It is as visible as the most remote storeroom in your neighborhood school. Ask to visit that room and you are likely to discover the entire language laboratory record library, several broken projectors that no one has the expertise or the desire to repair, a couple of yellowing teaching machine programs, several crumpled audio tapes and a stack of empty take-up reels, the laminating machine no one ever figured out how to use, a box of dried-out overhead projector crayons that would break if used, and maybe even the acoustic coupling for the computer terminal that had to be taken out when the budget was cut. Even operable and reasonably contemporary equipment is likely to show signs of disuse. Such storerooms, above all, do not need any more expensive dust-catchers. If electronic calculators are to be teaching aids at all, let them be out freely among the children.

References

Phillips, J. L., Jr. *The Origins of Intellect: Piaget's Theory.* San Francisco: W. H. Freeman and Sons, 1969

Rogers, J. J. "The Decline of Interest in the Language Laboratory as Seen in a Popular and a Professional Journal." *Notes and Abstracts in the History of Education* 22 (1969):13–14.

NEW PRODUCTS PROGRAMS PUBLICATIONS

• NEW PRODUCTS

Edited by **Lowell Leake, Jr.,** *University of Cincinnati, Cincinnati, Ohio:* **computing devices**

Thomas E. Rowan, *Montgomery County Public Schools, Rockville, Maryland:* **models, games, devices**

Albert P. Shulte, *Oakland Schools, Pontiac, Michigan:* **audio-visual materials**

INTRODUCTORY NOTE: The following article was reprinted with permission from *Mathematics Teaching*, March 1975, no. 70, pp. 42–43, copyright 1975 by the Association of Teachers of Mathematics [England]. It is a superb journal, and the cost is $8.00 a year for four issues, which includes membership in ATM. I have belonged to ATM for about six years and look forward to new copies of *Mathematics Teaching* as much as I do to publications of the NCTM. I urge readers to consider joining the ATM, or at least to have their schools subscribe to *Mathematics Teaching*. The address is Association of Teachers of Mathematics, Market Street, Chambers, Nelson, Lancashire, BB9 7LN, England.—LEAKE.

Specifications for Electronic Calculators

The day of the £10 [about $21] calculator has arrived. Maybe even this remark will seem to express unwonted surprise if the downward spiral in prices continues. But whatever the future holds there is no doubt that many teachers in schools of all types are beginning to consider buying one or more of these machines. With this in mind a group of interested teachers at the 1974 *ATM* Conferences set out to provide some guidelines which would specify what to look for in a basic machine. This is a machine which could be used almost universally in the classroom, with age groups of at least 5 to 16.

Perhaps a word of explanation is necessary here because this article is concerned with machines without memories and without scientific function keys like LOG, SIN, COS, etc. It is not that these machines are unimportant, but that with limited funds available a basic machine which meets nearly all the calculation requirements of children up to 16 would appear to have an overwhelming priority. The basic machine's price already matches that of the cheap transistor radio and there are numbers of cases on record where these have been bought for children by fond relatives. There are many apparently similar machines on the market and it was felt that some guidelines would be useful to get best value combined with maximum applicability from the money spent.

A basic machine should have the following, in order of preference.

1. Natural order arithmetic
2. Floating point
3. Underflow
4. Constant key to operate on all four operations
5. Eight digit display
6. Not too small, fingertip sized keys
7. Rechargeable batteries, with alternative mains operation
8. CLEAR ENTRY key

These items can be checked by using the following tests.

1. *Does it use natural order arithmetic?*

(a) Key in: 7 − 2 = (5)

Look for a separate "=" key.
Beware of machines that have "+=" or "−=" keys since these often demand an 'unnatural' arithmetic in the sense that calculations on the

machine do not follow the written order even with simple binary operations.

(b) Key in: 8 − 5 × 13 ÷ 7 = (5.5714285)

Each operator is a binary operator, working sequentially from the left. This calculation should be evaluated as though bracketed from the left; i. e. (((8 − 5) × 13) ÷ 7).

2. *Does it have full floating point arithmetic?*

(a) Key in: 123 ÷ 456 × 789 = (212.82236)

This will indicate whether the intermediate calculation result (0.2697368) is carrying maximum decimal places. This should be so, and even if a fixed decimal place setting is provided the intermediate calculation should still be carried out to the limit of the machine's capability. A severe fault in the decimal handling is evident if this chain calculation is not carried out satisfactorily.

Some machines demand that each part of a chain calculation be completed with an "=" key, and this may lose significant figures in the final result, because the displayed digits are only part of those used in a calculation. Depression of the "=" key may destroy the non-displayed digits.

(b) Key in: 1 ÷ 81 = (0.0123456)
and 1 ÷ 81 × 100 = (1.2345679)

If the machine has the capacity for doing arithmetic to more than eight digits the last operation will give the maximum set of significant digits that can be displayed on the eight-digit calculator. Clearly if the machine has this facility it will be more accurate, in the sense of avoiding rounding errors more successfully.

3. *Does it indicate when figures have been dropped?*

Key in: 88888888 × 2 =

The full answer is 177777776, with the decimal point indicating one digit too many for the eight-digit calculator. Observe how the machine handles this.

Some machines give 1.7777777, with the decimal point indicating one place dropped, associated sometimes with a flashing display. Other machines indicate this condition by the absence of a decimal point. This is less satisfactory because it appears to give an authentic result, whereas it is only a partially appropriate result. Normally all calculation should be terminated when overflow appears.

4. *How does the constant key operate?*
Almost all of the latest calculators have a constant facility, most usually indicated by a 'K' key, which allows entry of a given value to a store. The basic machine should allow this constant to operate on all four operations.

(a) Key in: CLEAR 2.3917 = K
Then 5.56 = ×

should produce 13.560939.

(b) Key in: CLEAR 2.3918 = K
Then 5.67 = +

should produce 8.0617(000).

(c) Key in: CLEAR 2.3917 = K
Then 5.67 = −

should produce 3.2783(000).

(d) Key in: CLEAR 2.3917 = K
Then 5.67 = ÷

should produce 2.3706986.

The clear key on many machines clears not only the display but also the contents of the 'constant' store. This store is best considered as holding a constant and an operator. On the same machines the operator needs to precede the constant. This is best checked using the multiplication example, 4(a).

5. *How many digits does the calculator display?*
There are a number of machines which can give more than the eight figures we feel sufficient. The important point to note is that eight figures is the maximum likely to be of use, so that more figures either cost more or alternatively are displayed in two parts. There seems no advantage in this. However, the apparent capacity of the machine may be limited to its displayed digits, in which case rounding errors may occur. The test at 2(b) is useful for deciding this.

6. *Is it large enough?*
If too small it can easily be knocked on the floor, or slipped into a pocket or briefcase. Keys need to be tested to see that fingers don't touch more than one key (inadvertently) at a time. The display should be clearly visible from a wide angle of view, clearly readable in daylight conditions.

7. *Can it be easily used in a classroom?*
This implies rechargeable batteries and a mains unit for recharging. The time between charges is also important. In general machines that keep all eight digits' positions lighted will drain the batteries more quickly. Light emitting diodes (leds) are more efficient than other forms of display. If a mains unit is bought for recharging check that it is rated at 240v AC. A number of those inspected are rated at 220v AC, and will have a diminished life on the mains supply in U.K. (*Ed. Note:* The last two sentences refer to the British electrical supply system and are of no interest to American readers unless you travel abroad and use a rechargeable or plug-in calculator made for use in the United States; in this

case you must remember that *many* countries do not operate on 110-volt AC.)

8. *Does it have a 'clear entry'* (CE) key? As already noted the clear key will clear both the display and the constant store, which may be undesirable in the middle of a calculation if a mistake has been made in entering a figure. A CLEAR ENTRY key allows the re-entering of a corrected number if the mistake is noted before the next operation key is pressed. This can be very frustating in a sequential calculation if no CE key is available.

These eight questions and checks can be put in the form of a matrix, and individual machines assessed comparatively. There is an order of priority here, although there may be rather less agreement over the order of items 6 to 8.

During this study it emerged that a number of machines already on the market came near to meeting the 'basic' specification. This is either because we (luckily) agree with some manufacturers, or that we have not been farsighted enough. It does however lead one to ask whether we should leave it to chance in the future? If the educational market is one of the large growth areas, then when are manufacturers going to take seriously the need to produce an adequate basic calculator at the right price? We have heard a figure of £5 [about $10] quoted. Who will take up the challenge?—*David Jesson, Division of Education, University of Sheffield and Frank Kurley, Rowlinson School, Sheffield.*

sharing teaching ideas

Calculators in the Classroom

In accordance with the National Council of Teachers of Mathematics' policy statement on the use of calculators in the classroom, our program for using calculators with our eighth grade was intended to reinforce learning and to motivate the learner in the study of mathematics. Calculators were used—

1. to provide continued and increased motivation toward the subject matter;
2. to furnish faster and more efficient ways of solving problems;
3. to allow problems of greater intricacy to be attempted; and
4. to contribute to further applications and exploration of related topics.

Fifteen calculators and adapters were purchased for one class. Initially, ten minutes of free time was set aside for students to explore the capabilities of the calculator. We had to overcome the fact that many students had access to calculators at home with different logics and keying sequences from the ones we had purchased. To offset this problem, a model calculator made of construction paper (50 cm × 100 cm) was employed to explain all the capabilities and limitations of the classroom machines.

Although many students claimed to be proficient in the use of a calculator, we found that at least two full class periods devoted strictly to problems in addition, subtraction, multiplication, and division were necessary to insure mastery of fundamentals. In particular, division problems proved to be most difficult for students to solve. For example, $0.3\overline{)432}$, $\frac{432}{0.3}$, or $432 \div 0.3$ all represent identical division problems, but on a calculator many students often confused the order of input. As a result, a great deal of practice in addition, subtraction, multiplication, and especially division was provided before students were asked to solve many applications.

The eighth-grade curriculum at the Bethlehem Central Middle School is composed of many diversified topics representing various branches of mathematics. Units of study covered during the school year include probability, statistics, proportion, percent, coordinate geometry, solid geometry, and algebra. Within each unit there exist several exercises involving the use of calculators that serve to reinforce learning activities from previous class periods.

By way of example, the volume of a rectangular prism and the volume of a cylinder were studied from the unit on solid geometry during one week of classes. In class on Friday, students took a short quiz reviewing the material on volume covered during the previous four days. In the remaining twenty to twenty-five minutes of class, students were required to complete an exercise with the aid of calculators. The exercise involved calculating the volume of two rectangular prisms and four cylinders. Students were given the dimensions of each figure, and formulas were written on the board for reference. The substance of that exercise is summarized in table 1.

TABLE 1

Figure	Length of Base	Width of Base	Height	*Volume = l × w × h
1. Rectangular Prism	19.85 ft.	11.3 ft.	7.7 ft.	1727.15 cu. ft.
2. Rectangular Prism	42.3 cm	39.6 cm	15.5 cm	25 963. 74 cm^3
	Radius		Height	Volume = $\pi r^2 h$
3. Cylinder**	7.4 in.		16.7 in.	2871.50 $in.^3$
4. Cylinder	29.5 cm		52.8 cm	144 280.48 cm^3
5. Cylinder	0.54 m		1.6 m	1.46 m^3
6. Cylinder	72 ft.		125.8 ft.	2 047 742.2 cu. ft.

* Students were instructed to round off all answers to the nearest hundredth place.
** $\pi = 3.14$.

The use of calculators in the classroom adds a distinctive feature to any mathematics program. On the basis of two years of experience, it is the feeling of the authors that calculators have a definite role in mathematics classes. We feel that calculators should be consistently used as a supplemental aid to learning.

The following list of guidelines is meant to assist readers who are in the process of initiating a program involving the use of calculators.

1. The purchase of all calculators should include a one-year warranty to replace or repair any malfunctioning machine.
2. Distinct and permanent identification is necessary for all calculators and adapters.
3. The authors highly recommend the use of electrical adapters as opposed to any type of recharging device. Adapters will provide uninterrupted and longer-lasting service from the calculators.
4. A locking cabinet must be provided to enhance the easy distribution, collection, and protection of all calculators and adapters.
5. Designated calculators should be assigned to students so that a particular machine is used by the same pair of students on a continuous basis.
6. Rules and regulations involving the use of calculators must be clearly stated and enforced so that students will exercise care in the operation of each calculator.
7. A trustworthy student should assist the teacher in the distribution and collection of calculators during a class period.
8. At least two full class periods of instruction should be provided to all students to learn the methods of operating a calculator.
9. Although educators should be urged to explore all avenues of incorporating calculator use into daily lessons, we highly recommend that the use of calculators not exceed one experience per week. The novelty of calculators in a classroom environment can easily be eroded by overuse; more important, basic computational skills might eventually become weaker.

Gregory Aidala
Peter Rosenfeld
Bethlehem Central Middle School
Delmar, NY 12054

Calculators in the Elementary Classroom

By **Paula M. Drake**

Should hand-held calculators be used in elementary school classrooms? The question has become a controversial issue and the subject of much discussion. I discussed the question with my sixth-grade class, and they have put the issue to rest. We have formed a Calculator Club.

After Christmas, children all over America came to class with Santa's newest electronic device, the pocket calculator. Many educators have frowned on the use of minicalculators in elementary classrooms because, they say, children have not yet learned the basic facts or developed the essential computational skills. It may be true that children in these grades have not learned basic number facts and have not yet developed skills in computation, but the hand-held calculator can be used as a motivational tool for the development of those skills and the learning of those facts.

My sixth-grade students use their calculators daily to check their work. When children have completed the work for the day and I have checked the work for concept attainment, the children take out their calculators. Most of the children have their own calculators, but those who do not are not at a loss. Many children are very willing to share their calculators with others in the class, or they offer to check others' work.

When the calculator check shows that work has been done accurately, the children experience an immediate sense of accomplishment. If the calculator check indicates an error, the children redo the example with the calculator. If the second calculator check confirms that there is an error in the student's work, then children who made the errors go over their own work to find their mistakes. When the children themselves have found the mistakes, they are more aware of the mistakes and can better avoid making similar mistakes in future examples.

If the calculator check is done by another student, then the calculator checker may take on the job of teacher when there are errors in the computation. The students may work together to find the error, which provides a natural opportunity for peer instruction. And all educators are aware of the tremendous value of peer instruction.

You might argue that minicalculators are all right in the sixth grade because sixth graders usually have learned basic facts and operations with numbers, but minicalculators have no place in the primary grades. I believe that hand-held calculators do have a place in the lower grades. Why not let the minicalculators serve the same purpose—let children check their own work with their calculators. Minicalculators will motivate children in any grade. Perhaps one calculator or those of children could be placed in a learning center entitled "Calculator Check-Out Center." When children check their own work, see the results immediately, and promptly correct their own errors, mathematics will come alive for them.

This immediate checking is not only a motivation for children; it also is a time-saver for the teacher. Teachers can spend their time helping children who have special needs. When children have errors, teachers may help them by asking, "Why was this problem

---

Paula Drake is currently teaching sixth grade at the Percy Hughes School in Syracuse, New York. She has taught grades one through six. Her teaching experience also includes teaching at the SUNY at Cortland Campus School, inservice courses in individualization in mathematics instruction and metrics, and work in the Teacher Corps Program at Syracuse University.

wrong?" If children can verbalize their errors, they will be more conscious of them. Calculators can also be used for drill. Children who are having trouble with basic number facts can be paired together; the calculator becomes the basic-facts "flash card."

At the sixth-grade level, I also use the twenty minutes before lunch each day to have a formal Calculator Club meeting. During such times we have discussed the *human* as a variable in the use of a mechanical device. If users do not have an idea of the answer they expect, then the calculator is of no value. We have also used this time to carefully read the directions included with the individual calculators. Most often this is never done if the calculator is in the category of a Christmas toy. A clear understanding of the calculator's functions and methods of operation raises the respect level from toy to educational tool.

Calculators today have almost become a household commodity because of their low prices and usefulness. Let's bring them into the classroom and make them an educational tool and a twentieth-century classroom commodity.

> Scene: A self-contained classroom. Twenty-six students are working on mathematics in an individualized setting. Out come the calculators, in come the television cameramen and news announcer.

This was the exact scene in my classroom recently. The children themselves convinced the television crew, and later the television audience, that the hand-held calculator has a place in the elementary classroom. The variety of ways this electronic device can be used in the mathematics classroom is boundless. Try it out and settle the issue for yourself. Your students will be the judges.

In my classroom, hand-held calculators have been a blessing in disguise.□

Minicalculators in the Classroom—What Do Teachers Think?

By **Henry B. A. Palmer**

Are attempts to introduce minicalculators in the classroom, particularly at the lower elementary grades, likely to be greeted with the same kinds of responses that previous innovations have encountered? Teachers were, and many still are, deeply concerned about the effects of such innovations as programmed instructional materials, computer-assisted instruction, contingency management, and behavior modification. Investigations of the reasons for their caution revealed the following: (1) erosion of the teacher's role—machines taking over teaching functions; (2) impersonalization of the teaching-learning process—reduction or elimination of the humane aspects of teaching and learning; (3) stifling of creativity—emphasis on precision, measurement, and mechanization inhibits creative expression.

This article seeks to open up debate and promote critical exploration relative to the use of minicalculators in the classroom. The idea originated from an assignment given to me by my superior, namely, to exert leadership in determining whether the advent of low-cost pocket calculators has discernible implications in mathematics instruction in grades K through 12 in the schools of California. In response to this assignment, I designed a questionnaire to survey the knowledge of all the leadership personnel responsible for mathematics instruction in Los Angeles County schools (K through 12) based on first-hand involvement and opinions. Some responses were also received from teachers and other leadership personnel outside Los Angeles County. The responses provide good indicators of the nature and extent of teacher concerns regarding the use of minicalculators in the classroom.

One of the biggest concerns of the teachers surveyed centered around the issue of how the use of minicalculators might affect students' computational skills. Some teachers seemed strongly convinced that students will not learn computational skills if calculators are available. This appears to be an important issue and must be resolved to the satisfaction of teachers if minicalculators are to receive widespread acceptance in the classroom. What is perhaps needed here is convincing evidence that calculators can be used in ways that do not adversely affect the development of computational skills.

Teachers' responses also reflected a high level of knowledge about the present status of hand-held calculators relative to the extent of use in the classroom, experimental programs, costs, reference materials, and other kinds of information. This is perhaps indicative of high interest in calculator use and further emphasizes the need for extensive discussion.

Although the survey showed a wide range of teacher reactions, on the whole teachers seemed favorable to the idea of calculators in the classroom. There appears to be general consensus that mathematics instruction should prepare students to cope with the real world, and since computers, calculators, and other mechanical devices are used regularly in commerce, business, and industry, their incorporation into the basic structure of mathematics curriculums seems necessary and appropriate. There was some concern, however, about the danger of promoting over-dependence on machines and concomitant dehumanization of education.

With respect to classroom management, student motivation, and academic achievement, teachers seem to view the use of calculators positively. Several studies were cited by the teachers that indicated significant gains in student achievement as a result of calculator use. Many stated that the use of minicalculators increased enthusiasm for mathematics on the part of both teachers and students, improved class attendance, motivated students to learn, provided concrete reinforcement of skills and concepts, and freed students from the tediousness of computation.

Other areas of the survey dealt with implications of using minicalculators in the classroom. Here again there was a wide range of responses; however, two issues seemed important to these teachers. One was ensuring equal access to calculators for all students. The teachers felt that inequality of opportunity could easily develop in those classroom situations where students provide their own calculators for doing assignments and taking tests. Some homes will not or cannot purchase such items for students. The other issue centered around the need for strong support from school administrators, both financially and educationally, so as to facilitate adequate teacher in-service programs and the acquisition, maintenance, and replacement of calculators. These teachers felt that there was a definite need for workshops to help them

A mathematics consultant for the office of the Los Angeles County Superintendent of Schools, Henry Palmer is responsible for computer-assisted instruction within the schools. He is also involved in training elementary and secondary school teachers through workshops and extension courses offered by local universities.

develop and improve their competence in the use of calculators in mathematics instruction. The identified areas of instructional needs included the following:

1. demonstration of how calculators can be used to make mathematics instruction more effective
2. identification of problem solving/integrated curriculum opportunities
3. demonstration of techniques for
4. development of evaluation designs to document effectiveness of calculators
5. methods of making assignments and writing tests that are geared to the calculator
6. information on software development

The results of this survey seem to indicate high interest among teachers in experimentation with calculators in the classroom. Therefore, it would seem appropriate to facilitate extensive regarding a wide range of issues relating to this subject.

Although many of the responses reflected varying degrees of reservation, no one was completely opposed to the use of calculators in the classroom. The general consensus appeared to be "We are living in the age of computers, calculators, and electronic devices. We must learn to utilize these facilities in the instructional process skillfully and effectively so that the present generation of learners will be adequately prepared to cope with the

ARE CALCULATORS A CRUTCH?

By ZALMAN USISKIN
The University of Chicago
Chicago, IL 60637

A major argument against the use of calculators is that calculators are a crutch. This argument underlies the thinking represented in quotes like the following:

> I understand the principle—get them motivated. But I have yet to be convinced that handing them a machine and teaching them how to push the button is the right approach. What do they do when the battery runs out? I see a lot of low-level math among college students who still don't understand multiplication and division. You take away their calculators and give them an exam in which they have to add 20 and 50, and they get it wrong. And I'm talking about business majors, the people who will soon be running my world. [James R. McKinney, professor of mathematics, California Polytechnic State University of Pomona, as quoted in the *New York Times*, Section IV, p. 7, 5 Jan. 1975]

The "crutch premise" is essentially that if you allow students to use a calculator for arithmetic problems that can be done by hand, then the students will be unable to do arithmetic when the calculator is absent. A corollary to this premise is that calculators should not be used with young students who are still learning arithmetic. Another corollary is that calculators should not be used with older students who have not yet learned arithmetic—that is, calculators should not be used in remedial classes.

If the crutch premise is accepted, then the presence of calculators will have no effect whatsoever on the arithmetic curriculum. Judging from the latest (1978) texts, it is a premise that is presently widely accepted. However, *the crutch premise is seriously open to question, both in its internal validity and in the validity of the conclusions that are reached from it.*

The crutch premise rests on the principle that a crutch is a bad thing. But in fact, for the injured person a crutch may be a good thing—even a necessity. In supermarkets and other stores, calculating cash registers are necessities because of their accuracy and speed. Furthermore, these store calculators came into wide use at a time when the general populace was taught at least as much about calculation as it is today. For both the injured man and the supermarket, the crutch has become a necessary tool. The capacity for a crutch (bad!) to be relabeled a tool (good!) extends to many situations, and many value judgments may simply depend on which label is perceived as applicable.

It is common to cite the case of a real or hypothetical student who takes a calculator into an exam only to have the battery run out, after which the student is helpless and confused. Such events do happen. But when they do, one must ask two questions. First, will the student allow this to happen on the next test? One would expect that a single experience of this kind would suffice and that a similar experience would be avoided. Second, for how many students in the same test was the calculator an asset? In short, one should be careful not to penalize the majority (those with calculators) in a test because of unwise decisions that are bound to be made by a few (those whose batteries run out).

When a computer or business machine breaks down in the real world, few organizations reject the idea of using the machine. Most get it quickly fixed, or they buy a new one. It is a fact of life that machines break down or are at times unavailable, but the increased level of performance that they make possible more than makes up for these inevitable problems. To cite an example closer to the student's world, if an

This article, with minor revision, is part of a position paper originally prepared by Zalman Usiskin and Max Bell entitled "Calculators and School Arithmetic: Some Perspectives," which was an appendix to "Electronic Hand Calculators: The Implications for Pre-College Education," a report prepared under NSF Grant No. EPP 75-16157. The opinions expressed herein do not necessarily reflect the position or policy of the National Science Foundation.

essay is required to be typed and the student's typewriter is not usable, the simple solution is to find another typewriter.

The crutch premise leads to the conclusion that calculators should not be used in remedial arithmetic classes (such as general mathematics courses in high schools or junior colleges). There is some reason to believe that this conclusion is false. Indeed, calculators may be more appropriate for these students than for any other students of arithmetic. Let us consider the present situation. In the National Assessment of Educational Progress (NAEP) studies, 34 percent of seventeen-year-olds incorrectly answered $1/2 + 1/3$; 26 percent incorrectly answered $1/2 \times 1/4$. These are substantial percentages, and they occur despite the existence of special courses in arithmetic for high school students in which such skills are covered. In fact, it is now estimated that about 40 percent of entering high school freshmen are not sufficiently adept at arithmetic to enter algebra. These students have had eight years of arithmetic and not had success at it. Though we have given them even more arithmetic (usually in general mathematics courses), the NAEP studies of seventeen-year-olds show that the total experience is still not successful. At present such people are condemned never to have the power to use arithmetic in their daily lives. For such people the calculator is not a crutch; it is the only way to get a right answer.

The crucial issue is not how the calculation is accomplished but rather knowing when and how to use arithmetic to solve problems or answer questions that in fact matter in the lives of people. In this sense, even the long division and multiplication algorithms are crutches to help us get answers. Insisting that all children must be excellent pencil-and-paper calculators puts the emphasis in the wrong place—on the means, rather than on the ends, of calculation. The ability to use the results of calculation is what we should expect from those who have completed their study of arithmetic.

Hand-held Calculators: Past, Present, and Future

Max Bell
Edward Esty
Joseph N. Payne
Marilyn N. Suydam

It should come as no surprise that the nonthematic essay in this yearbook is devoted to hand-held calculators. Surely no other device, whether designed as an aid to learning or not, has had more potential for influencing instruction in mathematics. No other topic is more timely for mathematics teachers at all levels. The position statement of the National Council of Teachers of Mathematics, adopted in September 1974, reflects this timeliness:

> With the decrease in cost of the minicalculator, its accessibility to students at all levels is increasing rapidly. Mathematics teachers should recognize the potential contribution of this calculator as a valuable instructional aid. In the classroom, the minicalculator should be used in imaginative ways to reinforce learning and to motivate the learner as he becomes proficient in mathematics.

As we think of the relation between calculators and precollege mathematics, it may be helpful to consider two futures: the immediate future (the one that is inexorably merging into the present) and the long-range future (which remains, at least for a while, comfortably distant).

Of course, there is no firm demarcation between the two: teachers, parents, and administrators are concerned with both. Some of the more pressing questions concern the immediate future: What can I do with calculators in my classroom tomorrow? Should I let my child use the family calculator for doing homework? Should I purchase a classroom set of calculators for my primary team, and if so, what kind should I get? But the long-range questions are there too: How will my fifth-grade year-long calculator program fit into the total mathematics education of my students? As a result of years of calculator usage, will my child be better prepared to deal with real-life mathematical problems or not? Rather than purchase a classroom set of calculators for next fall, should I wait a few years, hoping that prices will continue to drop while mathematical capabilities increase?

It is not the intent of this essay to answer questions like these, for there are no quick answers to give. Indeed, this sampling only hints at the complexity and depth of the issues involved. Instead, our purpose is simply to provide an overview of some recent reports that have dealt in whole or in part with calculators and to list some sources of information that might be helpful to the reader. To those ends, the next four sections are concerned with the report of the National Advisory Committee on Mathematical Education (NACOME), the Euclid Conference report, a report of a status study commissioned by the National Science Foundation, and the report from a calculator conference sponsored jointly by the National Institute of Education and the National Science Foundation. The fifth section is devoted to a potpourri of other calculator activities, and the last section summarizes some of the consistencies across sets of recommendations.

The NACOME Report

The National Advisory Committee on Mathematical Education was appointed by the Conference Board of the Mathematical Sciences in May 1974 to prepare an overview and analysis of school-level mathematical education in the United States—its objectives, current practices, and attainments. Funded by the National Science Foundation, the report of the the committee was released in November 1975.[1]

The topic of hand-held calculators appeared in four separate places in the report. In chapter 2 the strong trend to emphasize computation in the curriculum was severely questioned. The case for decreasing the emphasis on manipulative skills was seen as stronger than ever because of the impending universal availability of calculating equipment. Instead, emphasis was recommended for approximation, order of magnitude, and the interpretation of numerical data. For children who have not acquired functional levels of arithmetic computation by the end of grade 8, the committee suggested that they be provided with calculators to meet their arithmetic needs.

In a separate section on computers, the widespread availability of handheld calculators was viewed as a challenge to traditional instructional priorities. The following changes were envisioned by the committee (pp. 41–42):

1. The elementary school curriculum will be restructured to include much earlier introduction and greater emphasis on decimal fractions, with corresponding delay and de-emphasis of common fraction notation and algorithms.
2. While students will quickly discover decimals as they experiment with calculators, they will also encounter concepts and operations involving negative integers, exponents, square roots, scientific notation and large numbers—all commonly topics of junior high school instruction. These ideas will then be unavoidable topics of elementary school instruction.
3. Arithmetic proficiency has commonly been assumed as an unavoidable prerequisite to conceptual study and application of mathematical ideas. This practice has condemned many low achieving students to a succession of general mathematics courses that begin with and seldom progress beyond drill in arithmetic skills. Providing these students with calculators has the potential to open a rich new supply of important mathematical ideas for these students—including probability, statistics, functions, graphs, and coordinate geometry—at the same time breaking down self-defeating negative attitudes acquired through years of arithmetic failure.
4. For all students, availability of a calculator does not remove the necessity of analyzing problem situations to determine appropriate calculations and to interpret correctly the numerical results. The user must still determine which calculator buttons to push. With de-emphasis on the purely mechanical aspects of arithmetic comes an opportunity to pay close attention to other crucial aspects of the problem solving process and to treat more genuine problems with the "messy" calculations they inevitably involve. Facility in the mental estimation of arithmetic results, to check that one's calculator is functioning well and that correct problem analysis has preceded calculation, will continue to be useful.
5. Present standards of mathematical achievement will most certainly be invalidated in "calculator classes."

The committee also listed some questions it considered important to investigate through research (pp. 42–43):

> When and how should calculator use be introduced so that it does not block needed student understanding and skill in arithmetic operations and algorithms?
>
> Will ready access to calculators facilitate or discourage student memory of basic facts?

For which mathematical procedures is practice with step-by-step paper and pencil calculation essential to thorough understanding and retention?

What types of calculator design—machine logic and display—are optimal for various school uses?

How does calculator availability affect instructional emphasis, curriculum organization, and student learning styles in higher level secondary mathematics subjects like algebra, geometry, trigonometry, and calculus?

In closing the section, the committee expressed the view that calculators would allow students to feel the power of mathematics and would free time for concentration on the conceptual aspects of the subject.

In the final chapter, the following major recommendation was made (p. 138):

> That beginning no later than the end of the eighth grade, a calculator should be available for each mathematics student during each mathematics class. Each student should be permitted to use the calculator during all of his or her mathematical work including tests.

The report also identifies particular areas where "new curricular organizations, instructional materials, and courses are of urgent concern," recommends that "instructional materials at all levels" be developed for calculators and suggests that "curricular revision or reorganization in the light of the increasing significance of computers and calculators" is needed (p. 145).

The Euclid Conference

Within the past two years, the Basic Skills Group of the National Institute of Education (NIE) has begun to devote increasing attention to mathematics. It was natural that some of the first questions to arise should concern the nature of basic mathematical skills.

As a first step in sorting out some of the issues, NIE sponsored a Conference on Basic Mathematical Skills and Learning during October 1975 in Euclid, Ohio. Each of the thirty-three participants was asked to submit a position paper before the conference in response to the following questions:

1. What *are* basic mathematical skills and learning?
2. What are the major problems related to children's acquisition of basic mathematical skills and learning, and what role should the National Institute of Education play in addressing these problems?

The final report from the conference consists of two volumes—one containing the thirty-three position papers and the other containing a description of the background and organization of the conference, four working-group reports, and an essay by the conference cochairmen.[2] Not surprisingly, issues related to hand-held calculators were discussed in many of the position papers, in three of the four working groups, and in the essay.[3]

Calculators in the position papers

Although some of the authors mentioned calculators only in a single sentence, others devoted more extensive sections to the subject. The following partial list of topics is presented to indicate the kinds of issues that were treated:

1. The effect of calculators on the curriculum
2. Calculators used as an aid to early counting skills, deficiency in which may later lead to difficulties in addition and multiplication
3. Calculators as a vehicle for a reexamination of the placement and emphasis of various topics in the mathematics curriculum
4. Calculators as an aid to the teaching of programming in grades 5–8
5. The amount and kinds of paper-and-pencil algorithms now needed, both in and out of school
6. The importance of looking toward a future in which hand-held devices will be far more powerful and sophisticated than they are now

Calculators in the working-group reports

Three working groups discussed calculators. The first proposal from the group working on curriculum development and implementation dealt with the impact of calculators and computers. In particular, they recommended studies of—

> 1. alternative sequences for elementary instruction in arithmetic;
> 2. uses of the calculator as an aid and stimulus for arithmetic instruction;
> 3. the impact of calculator availability on problem-solving instruction;
> 4. the relative importance of various familiar fraction concepts in an environment of calculators (to include an investigation of curriculum topics in later courses such as algebra). [p. 11]

Further, they noted a need at the secondary level to "re-examine curriculum structures and priorities in light of increasing computer and calculator capabilities to perform traditional computations," saying that "this clearly affects the definition of basic skills" (p. 12).

The group working on goals for basic mathematical skills and learning noted that "the whole issue of the effect of the calculator on the teaching of arithmetic is a very complex one which deserves considerable investigation and consideration" (p. 18). The problem is one of finding the proper balance between a minimum of single-digit arithmetic and the amount of paper-and-pencil computation that is presently taught. The group concluded that it is a "question that needs further study and far more discussion among a broader base of people" (p. 18).[4]

The third working group to mention calculators was concerned with research priorities. It recommended as one of the "first priority questions of practical urgency" the following (p. 29):

> *True functional needs (remember Xerox).* In the midst of controversy about what skills people ought to have, this question calls for the matter to be studied empirically—track people's behavior and *see* where they need which skills. "Remember Xerox" reminds us that "needs" should not merely mean needs exercised by current practices. As with the office copier, there may be unrecognized needs. In particular, the potential of hand calculators must be explored. What are their innovative uses? Learning uses? What would be the long-range implications of substituting calculators for pencil-and-paper computation?

Calculators in the essay

Finally, one of the ten sections of the essay by Hilton and Rising was devoted to calculators. They cautioned against placing any premature restraints on calculator usage—restraints either on grade level or on machine logic—fearing that interesting avenues of research would otherwise be blocked. They noted a need for good curriculum materials, particularly to "support and extend conceptual understanding of mathematics and to facilitate the application of arithmetical techniques to the solution of real life problems" (p. 39).

The Report to NSF

In March 1975, the National Science Foundation, concerned about the potential impact of the calculator on the precollege mathematics curriculum, funded an investigation involving a critical analysis of the role of the calculator.[5] The study was designed to identify the range of beliefs and reactions about calculators and in particular the arguments that were being used to support positions strongly favorable and strongly negative toward the use of calculators in elementary and secondary schools.

Questionnaires were sent to teachers and other school personnel, to state supervisors of mathematics, to mathematics educators in colleges and universities, and to textbook publishers. In addition to asking for statements of arguments for and against the use of calculators, the form contained such questions as these:

How should calculators be used?

What uses are most important at various levels?

How should the curriculum be modified if calculators are readily available to students at all times?

What would you recommend to elementary and secondary school personnel considering the selection and use of calculators?

Appendix B of the report contains the complete sets of responses from questionnaires.

Articles in educational and noneducational journals and in newspapers, reports from calculator manufacturers, curriculum materials, position papers, conference reports, and other documents were surveyed for additional arguments. (Appendix A of the report contains an annotated list of these references.) Manufacturers were surveyed in an attempt to secure information on current and future sales and development. Research reports were checked to determine what had already been ascertained about the effect of using calculators.

In addition to the analysis in the final report, several position papers were prepared by persons who have devoted much thought to the promises and problems posed by the use of calculators in elementary and secondary schools. The position papers are incorporated as appendixes C, D, E, and F: appendix C, *Teaching Mathematics with the Hand-Held Calculator,* by George Immerzeel, Earl Ockenga, and John Tarr; appendix D, *Hand-Held Calculators and Potential Redesign of the School Mathematics Curriculum,* by H. O. Pollak; appendix E, *Some Suggestions for Needed Research on the Role of the Hand-Held Electronic Calculator in Relation to School Mathematics Curricula,* by J. F. Weaver; and appendix F, *Calculators and School Arithmetic: Some Perspectives,* by Zalman Usiskin and Max Bell.

The case for using calculators in schools

The analysis of positions that people hold regarding the use of calculators in schools made it apparent that there is much similarity in viewpoints. In addition, these same viewpoints are reflected over and over in published articles. The most frequently cited reasons *for* using calculators in schools were these:

1. They aid in computation. They are practical, convenient, and efficient. They remove drudgery and save time on tedious calculation. They are less frustrating, especially for low achievers. They encourage speed and accuracy.
2. They facilitate understanding and concept development.
3. They lessen the need for memorization, especially when used to reinforce basic facts and concepts with immediate feedback. They encourage estimation, approximation, and verification.

4. They help in problem solving. Problems can be more realistic, and the scope of problem solving can be enlarged.
5. They motivate. They encourage curiosity, positive attitudes, and independence.
6. They aid in exploring, understanding, and learning algorithmic processes.
7. They encourage discovery, exploration, and creativity.
8. They exist. They are here to stay in the real world; so we cannot ignore them.

The last reason—the pragmatic fact that calculators exist and that they are appearing in the hands of increasing numbers of students—is perhaps the most compelling. *How* they can be used to facilitate each of the other seven beliefs is therefore a question that must be answered.

The case against using calculators in schools

The most frequently cited reasons for *not* using calculators in schools were these:

1. They could be used as substitutes for developing computational skills: students might not be motivated to master basic facts and algorithms.
2. They are not available to all, and so some students are at a disadvantage.
3. They may give a false impression of what mathematics is. Mathematics may be equated to computation, performed without thinking. Emphasis is on the product rather than on the process; structure is de-emphasized. Mental laziness and too much dependence are encouraged; a lack of understanding is promoted. Some students and teachers will misuse them.
4. They are faddish. There has been little planning or research on their use in classrooms.
5. They lead to maintenance and security problems.

The first concern—that students will not learn basic mathematical skills—was one expressed most frequently by parents and by other members of the public, as reflected by newspaper articles. But few educators believe that children should use calculators in place of learning basic mathematical skills. Rather, they express a strong belief that calculators can help children develop and learn more mathematical skills and ideas than would be possible without the use of calculators. Much serious attention must be given by teachers and others to proving that this belief can be supported and become fact.

Empirical evidence

In his position paper on needed research, prepared for inclusion with the report, Weaver pointed out that—

> the very newness of calculators provides little of a research base upon which to build. . . . The extent of ongoing research is very difficult to assess; this also is true of the nature of that research. We are given hints from the brief progress reports released by some projects . . . but by and large we have precious little information—and none of it definitive—regarding the extent and nature of ongoing research. [p. 18]

Most of the studies to date have been exploratory. Some of the "hardest" data come from studies conducted by calculator manufacturers; not surprisingly, these indicate that students (*a*) can use the calculator with a variety of content and (*b*) achieve well when using the calculator. Many schools are checking data on their own students to determine the effect of the use of calculators: reports indicate that using calculators generally results in achievement at least as high as that which results when calculators are not used; in some instances, computation scores are significantly higher when calculators are used, and in others problem-solving scores are significantly higher.

A summary of suggestions for research that should be conducted to determine the potential and the problems of using calculators in schools includes investigations related to the following:

- When and how to introduce calculators
- Effective procedures for learning basic facts, computational skills, problem solving, and various mathematical ideas
- Effective algorithms for calculators
- Long-range effects of using calculator algorithms
- The need for paper-and-pencil algorithms
- The effect of using calculators with specific content and curricula, including the effect of curricular sequence and changes of emphasis
- The relationship between work with calculators and work with computers
- Changes in teacher-education curricula
- Optimal designs and functions of calculators

One very specific caution must be emphasized: attempts to restructure the curriculum, either extensively or minimally, *must not* proceed independently of research. The two are interwoven, and one cannot be effective without the other.

Recommendations for curriculum and instruction

A variety of recommendations were suggested by educators responding to the questionnaire. These recommendations ranged from the general to those supporting specific curriculum changes. For example:

1. Experiment and plan, finding meaningful ways to use calculators. Develop a school-wide policy and guidelines, incorporating calculators into the existing curriculum and developing new curricula as necessary. Plan a reasonable in-service program, evaluation, and research to support the use of calculators.
2. Survey available calculator models carefully, and buy good equipment commensurate with student needs. Make sure that all students have access to a calculator.
3. Think of calculators as tools to extend mathematical understanding and learning by making traditional work easier. The focus can be on process because the product is assured.
4. Change teaching emphases to concept development, algorithmic processes, applications of various operations, and problem solving using real-life and interdisciplinary applications. Place more emphasis on problem-solving strategies. Use practical, realistic, significant problems and more applications.
5. Do not ignore the development of computational skill. Spend less time on computational drill and more time on concepts and the meaning of operations. Use more laboratory activities where computation is involved but the emphasis is on learning mathematical concepts. Decrease the use of tedious, complicated algorithms; emphasize algorithmic learning, including the development of algorithms by students.
6. De-emphasize fractions and emphasize decimals, introducing them earlier.
7. Emphasize estimation and approximation (including mental computational skills), checking and feedback, exploration and discovery. Do more or earlier work with such ideas as place value, the decimal system, number theory, number patterns, sequences, limits, functions, iteration, statistics, probability, flowcharting, computer literacy, large numbers, negative numbers, scientific notation, data generation, and formula testing.

Position papers

The four position papers in the report present viewpoints indicating the need for both immediate and long-range planning. Immerzeel, Ockenga, and Tarr point out that to avoid "future shock" imaginative software must be developed. They make recommendations and provide a variety of specific illustrations for using the calculator at each of several levels, usually with topics from existing curricula.

Pollak describes two partial orderings that are often used in designing a mathematics curriculum. The first is dependent on the structure or nature of mathematics and may be called *content ordering.* A second partial ordering that must be considered in curriculum development is *societal ordering;* that is, topics in a mathematics curriculum are included and ordered according to the topic's worth to society.

If the introduction of a new device such as the hand-held calculator makes a significant change in either the content ordering or the societal ordering, then major curriculum modifications seem appropriate. For example, the algorithm for addition with decimals is the same as the algorithm for addition with whole numbers: the same buttons are pushed for each. Thus, it may no longer be necessary or desirable to delay the teaching of decimals until fifth grade.

A careful, extensive study of the impact of the calculator on the curriculum is needed: it seems that significant changes could or ought to be made. In their position paper, Usiskin and Bell present some initial suggestions on this task. They take exception to merely incorporating the calculator into existing curricula: "It is thus our belief that the insertion of calculators into K–6 classrooms using most existing curricula is fraught with peril" (p. 36). They argue for an alternative curriculum and provide an assessment of how the curriculum could be restructured. They note that those who view the present curriculum as optimally logical and sequential may find this threatening; their specific suggestions could, however, suggest to many teachers a different way of considering the use of calculators.

Closely related to both curriculum development and instructional concerns is the role of research. Weaver discusses some of the research questions that should have priority. He called attention to two points:

> The greatest thing we have to fear today about the calculator vis-á-vis school mathematics curricula is the degree of fear that already exists about the calculator vis-á-vis school mathematics curricula. [p. 2]
>
> To many persons the calculator threatens to violate certain tenets regarding school mathematics learning and instruction. . . . Sugestions for calculator uses are made within the constraints of those tenets . . . and any research that might be implicit in such suggestions would be similarly constrained. . . . Some other persons, however, appear to be willing—and possibly even anxious—to suggest calculator uses that may challenge certain of our cherished tenets. [p. 5]

Weaver distinguishes among three types of curricula—calculator-assisted, calculator-modulated, and calculator-based—and points out that "research should not be unmindful of such differential roles." He discusses six research questions for which answers should be sought; each in turn can lead to a series of investigations.

Conference on the Uses of Hand-held Calculators in Education

The Conference on the Uses of Hand-held Calculators in Education was convened in Washington, D.C., in June 1976 by the National Institute of Education and the National Science Foundation. Participants were charged to "produce a . . . planning document that will provide a well-defined framework for future research and development efforts . . . that NSF and NIE may use if they wish as a guide to program planning."

In addition to the conference participants, seven persons attended the conference briefly to make presentations. They ranged from representatives of the microelectronics industry with a rather intense "get with it lest the world leave you behind" message to a representative of a citizens group so disturbed about possible negative effects as to have already petitioned a state legislature for outright prohibition of calculators in schools.

The participants concluded early that calculators are certain to be a significant fact of life in the world outside of schools and that they overlap so heavily with what is done in school that schools may not be permitted any *choice* about taking account of them. That is, substantial influence of calculators on schools was seen as inevitable, for better or worse. With that conclusion, the work of the conference turned to the outlining of research and development to increase the probability of better rather than worse consequences.

Conference participants agreed that the role of the conference was not to try to *answer* questions about the effects of calculators in schools but rather to *identify* such questions and suggest initiatives in information gathering, research, and development. They were particularly concerned that a variety of questions be kept open for investigation that might otherwise be foreclosed by too hasty conclusions. They tried to sort out what initiatives in research and development might be fruitful in helping teachers, youngsters, and parents cope with the immediate challenges presented by calculators. But they also tried to identify ideas that might result in substantial new approaches to teaching mathematics.

The conference report summarized discussion about many aspects of present-day school mathematics and the opportunities and dangers presented by calculators. The recommendations that emerged from those discussions are synthesized here. For the specific recommendations, with their rationale and other details, a copy of the report should be obtained.

The recommendations are grouped into several broad areas of concern: dissemination, the development of an information base, the development of curriculum materials for the immediate future and for the long-range future, research and evaluation considerations, and teacher education.

Dissemination

A central information collection and dissemination center should be established to make easily available both samples of materials being produced and research reports related to the use of calculators in schools. The means to better dissemination of information, both to the public and to the teaching profession, should be explored.

Development of an information base

A set of studies should be commissioned to provide intensive and critical analyses of existing mathematics curriculum materials, current instructional practices, and actual uses of computation (and other mathematics). Such information would provide a basis for curriculum development and research efforts involving the calculator. In addition, calculator-relevant information related to different aspects of mathematics learning should be critically reviewed. A continuing study should monitor the calculator as an educational innovation.

Curriculum development for the immediate future

Many curriculum development and research efforts focused on particular topics and small components of courses or curricula should be undertaken both to fill immediately pressing needs in accommodating to calculators and to pioneer new work that may later become part of considerably restructured courses. Attention is directed in particular to the role of algorithms and algorithmic processes, problem solving, a variety of topics such as estimation, and topics not presently taught at particular levels.

Curriculum development for the long-range future

A variety of full-scale alternatives to current school mathematics programs, fully integrating the use of calculators, should be developed. Psychological and other theories or models should not be neglected in the development of these alternatives.

Research and evaluation considerations

Research and evaluation related to calculators must interact strongly with initiatives in curriculum development. Sampling and other aspects of research methodology should be given careful consideration. There should be development and research focused on new approaches to achievement testing and to assessing attitudes and other noncognitive outcomes of school mathematics instruction. Both achievement and attitude in relation to the use of calculators should be carefully assessed, including any change in the amount of time spent in mathematics instruction. Means should be found to assure truly critical evaluations of calculator-oriented materials (and calculators themselves) proposed for school use.

Teacher education

The training and retraining of teachers to help them respond to calculators and calculator-influenced curriculum materials must be made nearly equal in importance to the development of new initiatives aimed at school students.

Other Activities

Many schools are incorporating calculators into existing mathematics programs. Most of them seem to be proceeding with caution because of parental concern that calculators may have an adverse effect on achievement. To help allay these genuine concerns, schools are keeping rather careful records on achievement.

One specific example is from Livonia, Michigan. During the 1976–77 school year, five different fourth-grade classrooms were provided with hand-held calculators, five in each room. Each child had regular access to a calculator for about twenty minutes each week. It could be used in doing classwork or in other ways. Achievement data on a standardized mathematics test from these five classes will be compared with achievement data from five comparable control classes. If the results seem promising, the next question expected to be explored is, How can the calculators be integrated into the mathematics instruction in the classes? The proportion of class time devoted to calculators may be increased also if the data seem to support it.[7]

Also in Livonia, four units designed for use with hand-held calculators have been written in a self-teaching format. They are planned for use in a consumer mathematics course and a remedial mathematics course for students in grades 10, 11, and 12. If the pilot use of the units during the year seems promising, it is likely that five or six additional units will be written.

Another example, among many that could be cited, is from Western Springs, Illinois. There, with the aid of a parent advisory group, School District 101 is conducting a study in which the calculator is used throughout the district as a teaching aid at the fifth- and sixth-grade levels. The district is seeking the answers to five questions involving the effect of the use of the calculator on achievement, attitude, and motivation; how the calculator can best be used; and what curriculum changes are needed. In some fifth-grade rooms, each child has access to his or her own calculator, and highly directed units of study have been developed to use the calculator as the primary mode of instruction on whole numbers, fractions, decimals, and metric measurement. In other fifth-grade rooms, a limited number of calculators are available to use as supportive tools for checking paper-and-pencil computation. Some sixth-grade classes have five or more calculators for use in a nondirected, exploratory program. Several classes at both levels act as control groups. Testing indicated that the group using calculators with calculator-oriented materials scored significantly higher than the other three groups; the group using calculators in a supportive role scored higher than the remaining two groups. Implications for future efforts might be evident as data amasses.[8]

In general, certain patterns concerning the use of calculators are evident:

1. The district or school purchases a small number of calculators, which are given to teachers for exploratory activities. This is followed by discussion and decision on whether the district or school should purchase more (e.g., Macomb County, Mich.; Columbus, Ohio).
2. Remedial mathematics or Title I classes receive calculators for use with low achievers who have not previously learned computational skills well (e.g., Washington Irving High School, New York City; Berkeley, Calif.).
3. Calculators are placed in advanced science and mathematics classes in secondary schools (e.g., Lubbock, Tex.);
4. Pilot studies and research are being conducted on the effect of the use of calculators (e.g., with low achievers at the secondary level in Chicago; in schools in such California cities as Cupertino, Garden Grove, Los Angeles, San Diego, San Francisco, and Santa Barbara).
5. Teachers and students bring calculators into the classroom and use them when it seems feasible (instances too numerous to mention).

Among the variety of other activities, surveys of the attitudes of teachers toward calculators and of the uses being made of calculators have been conducted (e.g., in Philadelphia, Pa.; Shawnee Mission, Kans.; and the states of Ohio and California); these sometimes lead to the development of policy statements. Statewide conferences in, for instance, Michigan and Ohio have focused on the role of calculators; workshops have been presented at local, regional, and national mathematics meetings throughout the country. The National Council of Teachers of Mathematics has assembled in Reston, Virginia, a collection of calculators and materials for use with calculators and has developed a set of three films illustrating ways of using calculators.[9] In cooperation with the ERIC Center for Science,

Mathematics, and Environmental Education, NCTM has also prepared a compilation of teacher-suggested activities for use with calculators.[10] Several journals (e.g., the *Arithmetic Teacher*[11]) have had issues focused on the calculator. Some state professional organizations have produced collections of activities for use at various levels. For instance, the Iowa Council of Teachers of Mathematics has produced *The Hand-held Calculator,*[12] and the Michigan Council of Teachers of Mathematics prepared *Uses of the Calculator in School Mathematics, K–12.*[13] The interest and involvement of an increasing number of teachers, parents, and organizations is evident.

Summary

A variety of activities and recommendations have been reviewed. The intent of this overview is to inform the reader of some sources of information that are available: each cited document includes specific thoughts and ideas for action.

As the recommendations of various groups are analyzed, it is apparent that there is a high degree of overlapping. Across these groups, which include classroom teachers, supervisors, mathematics educators, mathematicians, parents, and others, there seems to be agreement that—

1. the calculator cannot be ignored: its use in the schools must be carefully explored;
2. the focus of attention must be on how the calculator can *best* be used to develop and reinforce mathematical skills and ideas;
3. studies must be made of the uses of mathematics needed by children and by adults, given widespread use of the calculator in our society;
4. research and curriculum development must proceed hand in hand;
5. although some attention must be given to immediate concerns, this must not preclude long-range planning for mathematics curricula and instructional practices that incorporate optimal use of the calculator.

In the task of evolving new materials, practices, and curricula, everyone who teaches and uses mathematics has a role to play. We are all learners in this process of adapting to change.

NOTES

1. Copies of the NACOME report, entitled *Overview and Analysis of School Mathematics: Grades K–12,* are available through the ERIC Document Reproduction

Service (EDRS), P.O. Box 190, Arlington, VA 22210. The October 1976 issue of the *Mathematics Teacher* contained a summary article as well as several individual reactions to the NACOME report.

2. Single copies of the Euclid Conference report are available free while the supply lasts from E. Esty, Mail Stop 7, NIE, 1200 Nineteenth St., NW, Washington, DC 20208. The report can also be obtained through EDRS.

3. For the convenience of those who have access to a copy of the Euclid Conference report, references to hand-held calculators are found on the following pages: volume 1: 5, 24–26, 45, 60, 76, 94, 101–4, 119, 127, 151–52, 159, 173, 175, 203–4, 213, 223–24; volume 2: 11–12, 17–18, 29, 38–39.

4. One professional group that has been studying these questions is the National Council of Supervisors of Mathematics. NCSM conducted its own conference in July 1976 as a follow-up to the Euclid Conference; its report will be available in April 1977.

5. The NSF report, by Marilyn N. Suydam, is *Electronic Hand Calculators: The Implications for Pre-College Education,* NSF Grant No. EPP 75-16757, February 1976. Copies of the report are available from EDRS.

6. Information on the report of the NIE/NSF conference can be obtained from E. Esty, Mail Stop 7, NIE, 1200 Nineteenth St., NW, Washington, DC 20208. The report can also be obtained through EDRS.

7. Information on the results can be obtained from Charles J. Zoet, Mathematics Coordinator, Livonia Schools, 15125 Farmington Rd., Livonia, MI 48154.

8. Further information can be obtained from Kay Nebel, Project Director, School District 101, 4225 Wolf Rd., Western Springs, IL 60558.

9. Further information on the films, including rental fees or purchase price, is available from the distributor, Encyclopedia Britannica Education Corporation, 425 N. Michigan Ave., Chicago, IL 60611.

10. The publication is available from the National Council of Teachers of Mathematics, 1906 Association Dr., Reston, VA 22091.

11. The November 1976 issue of the *Arithmetic Teacher* contained articles on a variety of calculator applications.

12. The publication is available for $2 from Ann Robinson, ICTM, 2712 Cedar Heights Dr., Cedar Falls, IA 50613.

13. The publication is available for $2 from Horace L. Mowrer, MCTM Publications Chairman, 2165 E. Maple Rd., Birmingham, MI 48008.

Calculators and the Curriculum

As calculators continue to decrease in cost, they become more readily available to students and teachers in the classroom and their effect on the curriculum gains greater attention among mathematics educators. Bell presents some guidelines for using calculators and asks some questions whose answers illustrate the need for new curriculum development. Additional questions are posed by Gawronski and Coblentz, who also offer some ideas for implementing the use of calculators in the mathematics curriculum.

The NCTM Instructional Affairs Committee has identified several meaningful uses of minicalculators to reinforce learning in mathematics. Hopkins proposes that we introduce calculators in the curriculum as early as kindergarten and then progress to more sophisticated ones as the need arises. Stultz agrees with Hopkins that calculators could be introduced early in a child's school experiences, and he suggests some specific areas of the curriculum where the calculator can be applied.

In a curriculum-related article, Guthrie and Wiles recommend a calculator tournament for students in grades 5 through 12. They also express their own position on calculators in the classroom. Teitelbaum suggests some specific ways that calculators can be used to advantage in teaching mathematics.

According to Schmalz, calculators are *in* and they are never going to be *out;* therefore, we cannot continue to teach mathematics in the same old way. Bell, in a second, more recent article in this chapter, presents a strong argument for using the calculator in secondary school mathematics. He cautions, however, that it will be necessary for individual teachers to be innovative and thoughtful in accommodating their classrooms to the new realities of a calculator age.

In the final article of this chapter, Pollak suggests rethinking the content and teaching of secondary school mathematics in the calculator era, poses some interesting questions, suggests several uses of calculators to improve the mathematics curriculum, and concludes by recommending a conference—patterned after the Cambridge conferences—to reconsider the school curriculum with the availability of calculators in mind.

Calculators in elementary schools? Some tentative guidelines and questions based on classroom experience

MAX S. BELL

Associate professor of education, University of Chicago

It seems certain that electronic calculators will cause something of a revolution in the availability and use of calculating power in the world outside of schools. This special issue of the *Arithmetic Teacher* indicates that there is considerable agreement that they should also play an important role in school instruction. But viable school roles will not be established without finding solutions to many problems: problems of philosophy, problems of curriculum and methodology, problems of design, and school management of the calculators themselves. In the belief that solutions to many of these problems should be worked out in actual classrooms, a small band of local teachers and myself began several years ago to explore classroom uses of calculators. This article reports some of our tentative conclusions and our questions resulting from these explorations.

As we began to think about these matters in late 1973, calculators cost five to ten times as much as at present and very little had been done with them in schools. With certain very welcome support[1] calculators were loaned to teachers nearby. (Batteries and repairs were also supplied, and this was probably essential.) The teachers used the calculators (usually two or three per classroom) in whatever way seemed fruitful to them and gave me notes and comments about their experiences. Obviously more systematic work than that now needs to be done, but the feedback from these simple trials provided much preliminary information, gave at least tentative answers to some questions, and suggested many problems to be investigated further.

It must be emphasized that the comments that follow are based on informal work in only about twenty classrooms, with some comments based on experience in only one or two classrooms. With that understood, it still seems useful to attempt here a first approximation to sorting out of some of the issues that many teachers will face soon in using calculators. Roughly speaking, the comments focus first on student reactions to calculators; second, on some pedagogical questions that may need further investigation; and third, on some practical and management questions.

STUDENT REACTIONS TO CALCULATORS

Is explicit instruction in use of calculators necessary?

Our experience indicates that from first grade on children learn to use calculators very quickly (usually within the first hour)

[1] These explorations were supported first by the University of Chicago and later as a small part of a National Science Foundation grant; neither, of course, is responsible for any of the opinions or conclusions stated here. I will try to use "we" in this article to represent a collective consensus and "I" to identify more personal opinions. At the risk of omission of some who helped, here are some of those most active in these explorations: Pamela Ames, Katherine Blackburn, Mary Ann Chory, Sherye Garmony, Nancy Harvey, Beverly B. Johnson, Kaye Letaw, Charles Nelson, Blythe Olshan, Mary Page, Patricia Ryan, Pauline Schaeffer, and Ann Wheeler.

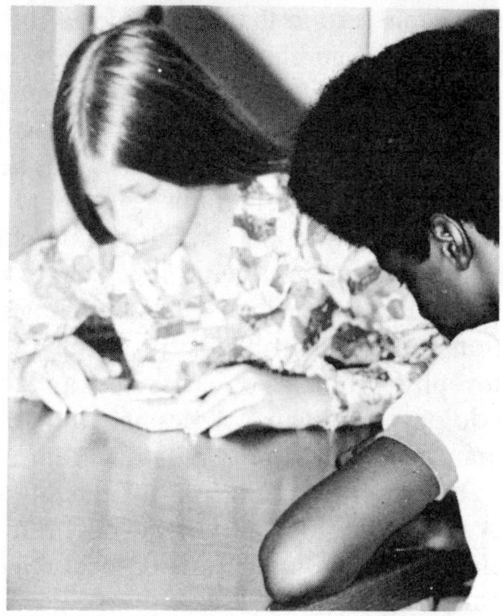

with at most a worksheet that confronts them with various possibilities appropriate to their grade levels. They tend to learn mostly about things they are already familiar with; for example if they don't know about division, they simply ignore that key. They learn both from the machine and from each other, with new wrinkles and shortcuts quickly shared. Hence, I advise against wasting time or money on booklets, filmstrips, cassettes, or whatever, that merely show "how to use a calculator." (Interesting problem material to give a context for calculator use is something else again, of course.) Also, I believe that children's capacities to ignore unknown keys without being distracted or to voluntarily forego use of certain keys when asked to do so indicate that teachers should not hesitate to use ordinary calculators "off the shelf" for exploratory work over a wide age range, not excluding first grade or earlier.

Are children interested in using calculators, and does the interest last?

There is almost invariably high initial interest, which persists over a long time *provided* students are given interesting things to do with the calculators. Indeed, they *demand* to be given things to do. Also, nearly all the teachers in these trials commented that "discipline problems" virtually disappear while calculators are in use, even in quite difficult situations. It could happen, of course, that as calculators become a very familiar part of our society student interest will diminish. But we have seen no evidence yet that "familiarity breeds contempt," either in classroom trials or for individual youngsters we know that have had calculators now for several years.

The main difficulty, of course, is that few existing school mathematics textbooks have really interesting problems that exploit the power given by calculators. That situation must be remedied, and the sooner the better.

Do children "naturally" detect errors and reject unreasonable results?

Even experienced operators sometimes get wrong answers, but they usually detect these errors from a habit of approximating or from the fact that calculator errors tend to be large ones. Similarly, we have found that children who have a good "number sense" can easily be led to use it in their calculator work. But most youngsters do *not* have that sure number sense and seem accustomed to accept whatever results they get, whether on paper or with a calculator. As to judgments about significant figures, nearly all youngsters merely write down whatever is on the calculator display, and often wish for even more digits. That is, the usual school neglect in teaching approximation skills and good sense about how many significant figures to retain is plainly revealed when children use calculators. Neglect of these matters must be corrected if calculators are used, but it should be corrected in any case.

Can calculators help in diagnosing gaps in conceptual understanding?

Some teachers in these trials found calculators helpful in identifying conceptual gaps, although systematic work was not done in this area. For example, using calculators in a decimal exercise with eighth graders revealed very quickly that some youngsters who had shown they were able to do the paper-pencil algorithms nevertheless understood very little about the *meaning* of the decimal notation. As another example, in a sixth-grade class the problem 38 ÷ 144 was variously answered (using a six-digit calculator) as 3.78947, 378947, 026388, and .26388. Each answer gives clear clues to appropriate follow-up.

I believe the diagnostic possibilities are promising and that further work should be done in this area.

Do children become curious about unfamiliar functions on the calculator?

As far as these trials go, this is still an open question. Third graders who knew a little bit about multiplication asked for more information about it in order to bring their own performance more nearly up to that of the machine. But in the same class no one asked about division and all ignored the divide key. After accidentally getting a decimal answer on the display, they asked about the meaning of decimals, but settled (temporarily) for a "whole number plus a little bit more" sort of answer. In other classes children unfamiliar with decimals in a general sense cheerfully accepted money-related explanations for decimal displays. Comparable results came from other grade levels.

Tentatively, the obvious answer also seems probable: valid discoveries from completely unguided exploration of unfamiliar keys are likely to be rare but there may be considerable potential in calculators for further exploration of partly familiar things. If, as we found, the unknown keys are not distracting, it should at least do no harm to keep further mathematical possibilities in the environment (by way of extra machine functions) and for certain youngsters and teachers some nice explorations might result.

SOME PEDAGOGICAL ISSUES

Do children become dependent on calculators? Does it matter?

Without denying altogether the possibility of overdependence as a long-range effect or as the result of unwise pedagogy, our trials to date indicate no such difficulties. Our experience indicates that children newly exposed to calculators begin by doing everything in sight on them but that they rather quickly gain good judgment about doing easy things in their head whenever they can. In any case, there appear to be easy safeguards against exclusive dependence on calculators; for example, most teachers in these trials periodically demanded paper-pencil work even with calculators present and students went along with this without resentment.

It seems unclear to me how much we should worry about possibly increased dependence on calculators. There are surely some things done only with great effort in school instruction that will be essentially irrelevant given easy access to calculators in the world outside schools. The "batteries may run down" argument for *no* change in how we teach calculation (or how much of it) seems silly to me. More troublesome to me is the fact that we know very little about how children learn mathematical concepts, and I suspect that for at least some children (perhaps only the "successful" ones) the very intricacy of arithmetic algorithms plus the patterns and rules that make them work may contribute to important, even if unspecified, learnings. Calculators can be used in teaching algorithms, of course, perhaps with the same benefits that some children get from paper-pencil algorithm work. Research in this area should surely be undertaken.

It should be easy for everyone to agree on one thing: with or without calculators it is crippling not to have good "reflexes" with respect to basic multiplication and addition results. Such reflexes are essential, for example, in making good estimates and estimates are as important with calculators as without them. For similar reasons it is also crippling not to have a sure feeling for effects of multiplication and division by powers of ten—a feeling that far too many people now fail to acquire. Some teachers in these trials used calculators effectively both in early work on "reflexes" and in promoting a sure feeling about operations with tens, hundreds, and so on.

Are there pedagogical consequences from choice of machine configuration?

I have already said that children in these trials adapted to a variety of machines without difficulty; they were also able to switch from one machine to another without confusion. But pedagogically speaking it is still possible that certain configurations have advantages over others. To discuss that, at least three aspects of configuration need identification:

(*a*) *Display.* Left to right or right to left entry; scientific notation or not; number of digits and capacity to alter that; size and color of display; and so on.

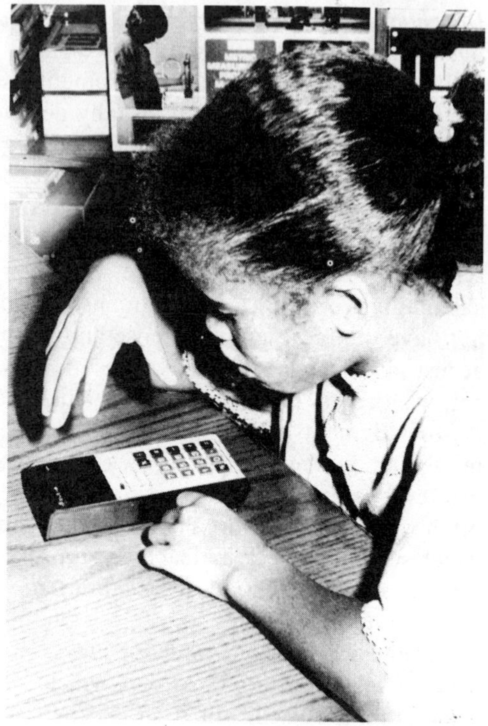

(*b*) *Keyboard.* What functions; number of functions; change-sign key; automatic or keyed constant; memories; multipurpose keys; programmable; and so on.

(*c*) *Type of logic.* Algebraic (equation) logic; arithmetic logic; commercial logic; Reverse Polish Notation (RPN) with a "stack."

These various aspects (plus power source) are combined in a bewildering array of choices in the marketplace. A consumer-oriented machine has now emerged that is virtually identical among major distributors and is quite inexpensive. Pending fuller investigation of what would be "ideal" for school use, this inexpensive standard calculator seems quite adequate for many school uses.

The possible pedagogical consequences from various combinations of display, keyboard, and logic are far from clear at present and should be investigated. Such investigation may turn up nontrivial consequences and if so, such finding can influence design of calculators for the school market. Listed in items 1 through 9 below are some questions or conjectures based on our experience that suggest further investigation or creation of specific cirriculum or pedagogical sequences.

1. Very limited work with first-grade children suggests the possibility that the standard display that feeds in digits from the right may lead to more number reversal errors in writing or reading numerals than displays that feed in from the left. (That is, for 78,

first

then

on most calculators, versus

and then

on some calculators.) This needs more careful investigation.

2. Displays necessarily have limits on number of digit places and children seem to find that dismaying, even though the usual eight-digit display shows numbers larger than they normally see or use. But there are obvious lessons to be learned from such display limitations and ways of coping with them. Ways to teach those lessons need to be devised.

3. When decimal answers are involved, the standard eight-digit display often gives much more "accuracy" than is warranted and there are valuable lessons to be learned from this. But for younger children so many decimal places is often merely a confusing overload of information. Both of them and others one can imagine pedagogical advantages to the ability to choose how many decimal digits will be displayed, especially if one can shift the display several times (with automatic round off) for a given result, as is possible with some calculators. Indeed, just the necessity to decide in advance how much decimal information to keep would teach something useful.

4. Children are sometimes dismayed by the penchant of calculators to give decimal answers where there "should" be whole numbers; for example, 1 [÷] 3 [×] 3 [=] **0.9999999**. A few calculators cleverly avoid such "difficulties," and presumably it would be possible to program a calculator with an integer arithmetic option. But perhaps the lessons of round-off error are too valuable in themselves to try to overcome the "wrongness" of the standard machine. How careful to be in this respect may be a function of the age of the child, but then again it may not. We should investigate these matters.

5. Most division problems on a calculator give decimal answers, but there are many real-life situations leading to division where quotient and remainder is the appropriate response, or where quotient with fraction is easier to interpret. (For example, if I divide 13 cookies among 4 people, ...) Remainders (and hence fractions) can be retrieved from the decimal answer and doing so might at some point be a valuable lesson, but perhaps not for children first learning about division. Perhaps calculators could have a "Q" or "(Q, R)" option switch, perhaps set by the teacher in early grades and later a student choice depending on the problem situation.

6. Nearly all the printed advice I have seen on buying calculators for school use specifies algebraic (that is equation) logic, but that could turn out to be poor advice from the standpoint of pedagogy. For example, it is far from certain that equation formats ($7 + 8 = \square$) provide a more "natural" way for children to think than the column formats suggested by RPN logic (enter the numbers, then operate). Some of our work with first-grade children suggests the opposite conclusion. Also, with equation format parentheses are sometimes needed (some calculators have them) or elaborate rearrangement of the calculations is sometimes needed, or one must sometimes record intermediate results on paper

or in calculator memory. There are useful learnings in all those things that must be balanced against the power with RPN logic (with a "stack") to usually do without parentheses, rearrangements, or recording of intermediate results. Also, we might investigate whether algebraic or RPN logic is more easily transferable to thinking about computer programming. This is not to suggest one over another choice of machine logic but merely to say that our experience indicates that the choices are not so obvious as is often suggested by printed advice and that there may be a rich field here for pedagogical inquiry.

7. We have already suggested that interesting pedagogical inquiries could follow from having machines available with more functions than are immediately familiar to a child. Might a third grader be led to ask about square roots? (If so, they could be easily explained with a calculator at hand.) What about base-10 logarithms, say at a time when integer powers of ten and their link to our numeration system is already well in hand? (With a calculator and hence no nonsense about extrapolation from tables, and so on, the discussion might be manageable much earlier than is now the case.) Would just the fact of seeing a variety of function names daily, even if never used or explained, make children more receptive to work with them later on?

8. In the typical calculator, all input and intermediate data is lost unless recorded by hand. Printing calculators keep track of all that, so are preferred by many for school use. But keeping a written record may be valuable in itself. It may also decrease the remoteness and mysteriousness of the "answer machine." Many children would have to be carefully taught to keep such records. Every teacher is familiar with children who regularly erase all the work they do and record only final answers, indicating that "neatness" is taught to many children as a virtue that takes precedence over demonstration of process. Again, the role of record keeping in calculation may merit investigation.

9. Scientific notation as available on some calculators might seem at first glance to be too complicated a code for use with young children. But our first reaction may be faulty and we should at least check on it by investigating when and how children can learn to use such a code with understanding. Whenever it is teachable, it also helps teach important things about our number system and opens up many new possibilities for genuine applications of arthmetic.

SOME COMMENTS ON MANAGEMENT PROBLEMS

Are calculators durable enough for classroom use?

As Joy Rogers points out in this issue, the most successful teaching aids are those tough enough for direct use by children. Also, since teachers usually lack time or energy or funds to attend to repairs, whatever gets broken is simply no longer used.

With calculators, a few new machines usually fail fairly soon, and vendors exchange these immediately. We have found that dropping and rough handling by children is rare, and that in any case being dropped seldom hurts the calculator. But there are occasional failures. Our experience so far indicates that if a school is to furnish calculators, there should be steady use for the first few hours to weed out and exchange initial failures and careful monitoring during the warranty period (usually one year). Thereafter, one might plan for about a 10% to 20% per year replacement (or repair) rate. With inexpensive calculators that should not prove a serious barrier to use.

What about losses from thefts?

For many teachers and administrators this question represents one of the major barriers to school use of calculators. One teacher flatly refused to try out our calculators with the comment "Those things have legs!" We assured the teachers in these trials that part of the experiment was to assess both durability and risk of theft, that they should exercise ordinary prudence, and that they would not be held responsible for losses. We lent about fifty calculators in a wide variety of situations with few special precautions. As nearly as we can tell, none were stolen by children actually using them but in a one-year period about five were lost in after-hours thefts from desks or cabinets. (We lost at least as many in loosely controlled lending to university students.) Lately we have etched numbers on the calculators, made up boxes with numbered storage cells (so it is obvious when one is missing) and put the boxes in locked closets overnight. This has so far prevented further losses.

If schools are to provide calculators, prudent planning would allow for the probability of some thefts and other losses. With calculators costing less than $10 this seems manageable. But many problems, including that of loss or theft, are avoided as far as schools are concerned if children own (or rent) their own calculators and I believe that alternative deserves serious consideration. For one thing, school ownership very often means restriction to classroom use only, both to save money by multiple use and because possible problems (to youngsters themselves) from loss, breakage, or extortion may make teachers unwilling to issue school-owned calculators for out-of-class use. Yet it is surely true that we will get the most out of calculators if students use them not only in mathematics classes but also at home, in other classes, while shopping, or whatever.

What source of power is best?

If schools are to buy calculators, decisions among penlight or transistor batteries, rechargeable batteries, or adapters are hard to make and I know of no fully satisfactory solution. Most printed advice on the subject advocates rechargeable calculators, but I'm not so sure. The extra cost is nontrivial. To plug in a classroom set for recharging is a nuisance. Existing rechargeable batteries have some quirks; for example, some acquire a "memory" for undercharging that keeps them from accepting a full charge.

On the other hand, keeping classroom sets of calculators supplied with ordinary batteries is a major nuisance and expense. While a calculator with small display and nine volt transistor battery is usable for several months by a single person exercising some care, the battery in that same calculator in continuous classroom use may last only a few days. Plug in adapters (not battery rechargers) are available for battery-operated machines, but few classrooms are equipped with enough electrical outlets to accommodate a classroom set, and portability is severely restricted. Adapters will often accommodate several machines by adding extra pigtails, but few teachers can do the required wiring.

Our experience indicates that problems in this area are a major deterrent to routine classroom use of calculators; especially use of school-owned sets. Again, many such problems are avoided as far as schools are concerned with individual ownership or rental since each student makes sure his own calculator has good batteries or is fully charged. We should expect that battery and calculator manufacturers will work on giving us better solutions than now exist. But I hope these problems do not deter widespread classroom trial and experimentation with calculators; the payoffs are more than worth the difficulties.

SOME CLOSING REMARKS

Our experience as partly outlined in this article plus the impact calculators are bound to have on the world outside of school leaves me with no doubt that over the next decade calculators should come to play a prominent role in elementary as well as secondary schooling. If this happens in a sensible and orderly way, then mathematics education a decade hence could look very different than it does now, and the ability of people to actually make use of mathematics in their jobs and everyday lives could be substantially enhanced. Orderly development would include in the first instance a great many teachers beginning soon to experiment with using calculators in their own ways with their own classes, even if only with one or two calculators. It would include establishing mechanisms for providing accurate consumer information and rapid review to help school people cope with what may be a flood of "hardware" and "software" (books, films, and so on)—some excellent, some mediocre, some rip-offs. It would include, I believe, fundamentally new curriculum development initiatives that would, among other things, enrich the calculation and number content of early schooling with greatly increased use of concrete materials, with calculators, and with problem material rich enough and interesting enough to give children some reasons to *want* to deal with numbers. It would include orderly investigation of some of the questions raised earlier in this article and in some of the other articles in this issue. The list could obviously be extended; the important thing is to try to nurture the fundamentally new possibilities offered by calculators with more patience and care than is often the case as educational fashions and supposed panaceas come and go.

Calculators and the mathematics curriculum

JANE DONNELLY GAWRONSKI
and DWIGHT COBLENTZ

Curriculum coordinators for the San Diego (California) County Department of Education, Jane Gawronski and Dwight Coblentz have been actively involved in the introduction and development of calculator-assisted, problem-solving activities in the mathematics curriculum.

Electronic calculators are small, practically silent, accurate, easy to use, and very fast. In addition, they have become virtually ubiquitous. Shoppers in supermarkets, adults balancing checkbooks, and students in the classroom have all been seen using calculators to help solve a problem. Calculators may even replace the transistor radio as a status symbol.

Some calculators can perform many complicated functions but most elementary or middle school students will have little need for one that does more than the four basic operations, square roots, percentage, and possibly powers. Most of these students will have highest need for a calculator to perform arithmetic calculations of addition, subtraction, multiplication, and division. These machines are typically called four-function calculators, may be hand-held, and have an electronic display. Calculators of this type can now be purchased for less than $20, and students from elementary school to graduate school are purchasing them or receiving them as gifts. Approximately 12 to 15 million hand-held calculators were sold in the United States in 1974 (Grosswirth 1975) and sales have not diminished since then. It is interesting to note that a president of one of the manufacturers commented that "One of the best selling seasons is the back-to-school season" (Grosswirth 1975). The low cost and availability of calculators, however, is not sufficient justification for using them as an instructional tool. We need to know if their use will enhance the understanding and achievement of the objectives of the mathematics program currently being taught or if they will precipitate a change in these existing objectives.

Etlinger (1974) has characterized two differing views on the use of the calculators, a "functional" view and a "pedagogical" view. In the purely functional view, the calculator is considered as a device much like an eraser or classroom desk—a device that can do the chores involved in tedious arithmetic computation, thus saving time and frustration. The other view, a pedagogical one, looks at the calculator much like a textbook, flashcards, or a manipulative device. The use of the calculator in this view does not replace something—rather it is another means to facilitate learning.

This distinction is certainly not a

good/bad distinction. Both uses could be good or bad depending on the task to be performed and the age of the student. High school students may be using the calculator in a purely functional way when they use it to determine the square root of a number needed to solve a quadratic equation. Yet, sixth graders might be using a calculator in a purely pedagogical sense when they attempt to determine a method or algorithm for obtaining the square root of a number. In each of these cases the task is to find the square root but because of the purpose or motivation and level of students, the one use is functional and the other use is pedagogical. When considering the justification of the use of a calculator in the curriculum, both views or positions can be rationalized.

Problems begin to arise when we examine more closely these two uses of the calculator. When considering the calculator as a functional tool and recognizing that it can do all of your arithmetic computations for you, do you in fact have to learn the algorithms for carrying out these operations—particularly those dealing with multiplication and division of three- and four-digit numbers. The calculator would change the current objectives of mathematics programs if we agreed that the calculator should take over and always be used to perform these operations. Instead of learning to multiply two three-digit numbers by using an algorithm, the new objective would be to find the product of three-digit numbers with the use of a calculator.

We cannot agree with this later premise if it means that one does not have to understand the nature of the operation if a mechanical means of obtaining the result is available. Relying on the machine for all calculations leads to depending on its use for problems that can be and should be done mentally. For example, $25 \times 2 = ?$; $10 \times 10 = ?$; $\sqrt{25} = ?$; $4 \times 4 = ?$. Furthermore, an understanding of the process may give added insight into its application in problem solving.

There exists a minimal level of computational skill for most people that should be at their command without mechanical support. For a small percentage of the population this minimal skill level might be low enough for the calculator to be used for nearly all arithmetic operations. Whenever these individuals are identified, usually in the upper elementary grades, they should be provided with a calculator to help them with all arithmetic computations. We are not suggesting that the machines be denied the other students, but the teacher should have a clear understanding of the purpose of the use of the calculator as related to the ability of the student.

All students will profit from having access to a calculator. If handled properly, the calculators will help a student develop a better understanding of the algorithms. We had not realized the extent to which some students confuse the algorithm with the abstract notion of an operation until a fourth grader was observed using a calculator to help him check the multiplication of two three-digit numbers. He was observed first working this problem on paper.

$$\begin{array}{r} 324 \\ \times\ 165 \\ \hline 1620 \\ 19440 \\ 32400 \\ \hline 53460 \end{array}$$

He proceeded correctly—first multiplying by 5, obtaining 1620, then by 60, and so on. The algorithm was carried out correctly. He was then asked to check the answer with his calculator. He proceeded to multiply 5×324 and recorded this on his paper. Then he multiplied 60×324 and recorded this on his paper, then 100×324 and recorded the answer. He then added the three answers he had recorded. Multiplication to this student was the algorithm. He had lost sight of the fact that multiplication is a binomial operation assigning a real number to a pair of real numbers, and the machine could make this assignment for him. We may discover many more interesting thought patterns by observing students working with their machines.

There is no doubt that the use of a calculator can save time in solving problems when number calculations are involved. It actually relieves the drudgery of computation. Teachers also report that there is a high level of motivation by children to use a calculator. Even young children themselves will initiate activities with a calculator—pressing buttons to see what happens. Eventually, some discover how to use a calculator by doing problems they can check. Children seem to want to test the machine and have been observed working problems on scrap paper to see if the machine is correct. This high level of interest in using a calculator seems to be maintained. At least, informal observations of teachers indicate that children are interested and motivated and that a high level of motivation is maintained.

Calculators are also particularly useful for more able students since they can use a calculator as a tool (functional use) with skills they have already mastered. This might even provide for their advancing more rapidly (pedagogical use) on the mathematics curriculum. On the other hand, the calculator might also aid the less able student in grasping mathematical concepts (pedagogical use).

Tedious and unnecessary calculations can consume time and destroy a child's interest in mathematics. When the calculator is used to eliminate this drudgery, more time will be available to focus on problem-solving skills. The idea of a basic skill may need to be reexamined. Universal use of a calculator may shrink the curriculum in some ways but it will certainly expand the curriculum in a problem-solving direction. The calculator is not going to replace the basic skills of mathematics and develop a generation of machine-dependent learners. But, when it comes to the actual calculation of the product of a three-digit number and four-digit number, most of us will probably reach for a calculator—not paper and pencil.

There are questions and concerns that need to be considered. Some of these are more of a philosophical nature and others are of a research nature. For example, there is the concern of who should supply the calculators—schools and school districts, or individual students. There is the potential here for students with an economic advantage to be buying an academic advantage. This is a problem from the elementary school to the graduate school level.

Questions of a more researchable nature include whether the immediate feedback from the calculator facilitates or hinders the development of the ideas behind the arithmetic operation. Also of concern is whether a child will think about more or fewer different methods to solve a problem when a calculator is available. These are questions that need to be investigated as the role of the calculator in the classroom evolves. The calculator has the potential for becoming a viable instructional tool for problem-solving activities and computational skill development. There are some researchable questions to be examined as well as curriculum uses to be identified. At the present we are convinced that calculators can provide another strategy for helping children to think, create, and learn mathematics.

References

Etlinger, Leonard. "The Electronic Calculator: A New Trend in School Mathematics." Educational Technology 14 (December 1974):43–45.

Grosswirth, Marvin. "Calculators in the Classroom." *Datamation* 21 (March 1975):90–95.

Minicalculators in Schools

NCTM Instructional Affairs Committee

With the decrease in cost of the minicalculator, its accessibility to students at all levels is increasing rapidly. Mathematics teachers should recognize the potential contribution of this calculator as a valuable instructional aid. In the classroom, the minicalculator should be used in imaginative ways to reinforce learning and to motivate the learner as he becomes proficient in mathematics.

The Instructional Affairs Committee has identified the following justifications for using the hand-held calculator in the schools. To support and illustrate these statements, problems from a wide range of grade levels are also included. It is hoped that this limited selection will show that the minicalculator can be a useful tool in every classroom.

The minicalculator can be used to encourage students to be inquisitive and creative as they experiment with mathematical ideas.

1. Find the solution in the tale about the man who invented the game of chess and asked to be rewarded in wheat. The rate of payment would be 1 kernel of wheat for the first square, 2 kernels for the second, 4 kernels for the third, 8 for the next, and so on. How many kernels would he receive for the 64th square? What is the total number of kernels he would receive?
2. Solve the following equation by the trial-and-error method.

$$\square^3 = 117\ 649$$

3. Use your calculator to assign a decimal fraction to the shaded part in each of the following situations.

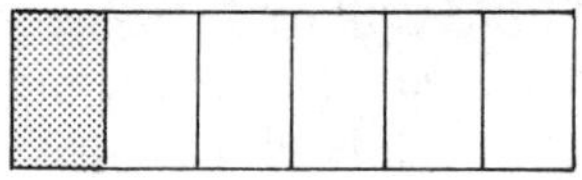

____

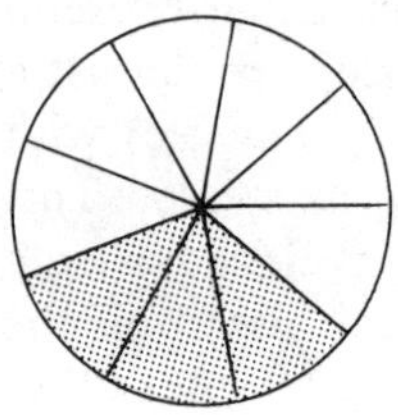

____

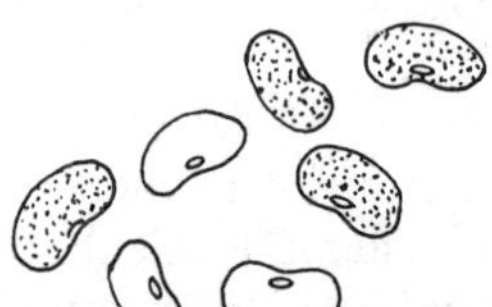

____

The minicalculator can be used to assist the individual to become a wiser consumer.

1. Find the unit price of various sizes of containers of a product (such as detergent) and compare the values to determine the best buy.
2. Find how many years it would take to double an investment of $5000 at an interest rate of 7 1/2% compounded annually, quarterly, and daily.
3. Find the actual cost of a bicycle if you buy it by making 12 installment payments of $9.67. How much do you save if you pay the cash price of $99.95?

This report was approved by the NCTM Board of Directors as submitted by the Instructional Affairs Committee. This committee currently consists of the following members: Jerry A. McIntosh, chairman; Joseph R. Caravella; Gloria F. Cox; Shirley M. Frye; Judith E. Jacobs; Jane E. Martin; Alan D. Nicholson; Tom Denmark; Laverne McMillan; and Harold D. Taylor. Former members who assisted with the preparation of this report and whose terms ended in April 1975 were C. William Engel, George Immerzeel, and Dorothea I. Peeler.

The minicalculator can be used to reinforce the learning of the basic number facts and properties in addition, subtraction, multiplication, and division.

1. Estimate the sum of these numbers:

$$\begin{array}{r} 3422 \\ 7958 \\ 5284 \\ 6331 \\ 8491 \\ +1206 \\ \hline \end{array}$$

   Find the sum using your calculator. Compare the results.

2. Circle the expressions that are equal to the *key* expression in each group. Use your calculator if necessary.

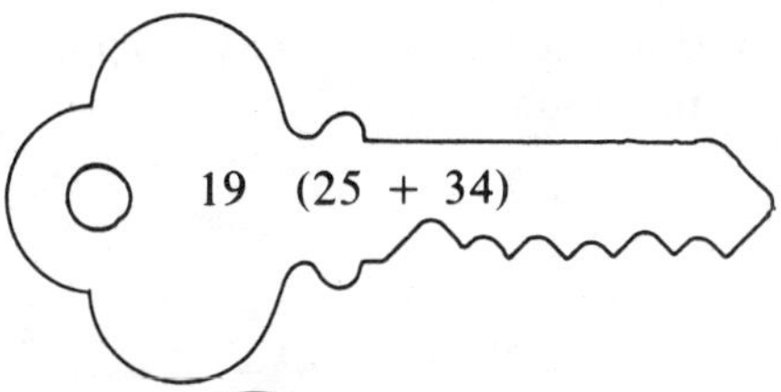

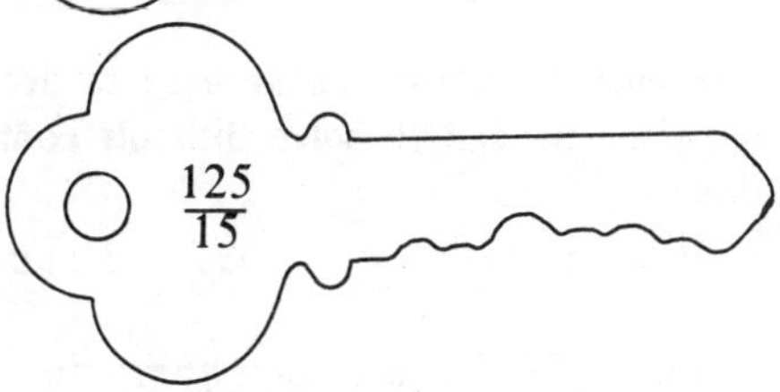

(19 × 25) + 34	25 (19 + 34)
(19 × 25) + (19 × 34)	(25 + 34) 19
19 × 59	19 (34 + 25)
$\frac{125 \times 6}{15 \times 6}$	$\frac{1250}{150}$
$\frac{125 + 10}{15 + 10}$	$\frac{12.5}{15}$

3. Use the minicalculator as a mechanical flash card when pairs of students practice their basic facts. One partner calls out a basic fact, such as 7 × 6, and computes the product on the calculator. His partner writes the answers or gives them orally. The student with the calculator checks the answers. Then they switch tasks.

The minicalculator can be used to develop the understanding of computational algorithms by repeated operations.

1. Add 528 seven times. Compare this sum with the product of 7 × 528.
2. Find how many times 14 can be subtracted from 1358. Check this by division, (1358 ÷ 14). Without using the division key on your calculator, solve 3960 ÷ 264.
3. Use your calculator to find the following numbers:
   (*a*) a number you can multiply times itself to get 5329 as a product
   (*b*) a number you can divide into 123 and have a remainder of 6

The minicalculator can be used to serve as a flexible "answer key" to verify the results of computation.

1. Use your calculator to check these computations. Circle the correct answers.

$$\begin{array}{r} 2016 \\ \times\ 31.45 \\ \hline 63403.20 \end{array} \qquad \begin{array}{r} 61.49 \\ \times\ 120.9 \\ \hline 7432.93 \end{array}$$

$$\begin{array}{r} 12.53 \\ 496\overline{)\ 6214.88} \end{array} \qquad \begin{array}{r} 2.45 \\ 296\overline{)\ 725.2} \end{array}$$

2. Place the decimal point in each answer of the following problems. Use your machine to check your answers.

$$\begin{array}{r} 35 \\ \times\ 1.6 \\ \hline 560 \end{array} \qquad \begin{array}{r} 5.49 \\ \times\ 3.96 \\ \hline 2174 \end{array} \qquad \begin{array}{r} 9.08 \\ \times\ .035 \\ \hline 31 \end{array}$$

$$\begin{array}{r} 259 \\ 25\overline{)\ 64.83} \end{array} \qquad \begin{array}{r} 975 \\ 1.3\overline{)\ 12.685} \end{array}$$

$$\begin{array}{r} 867 \\ .54\overline{)\ 4.684} \end{array} \qquad \begin{array}{r} 97357 \\ .07\overline{)\ 68.15} \end{array}$$

3. Use your calculator to check whether the following statements are correct:
   (*a*) If I buy 8 cans of beans at 37 cents a can, the total cost will be $2.96.
   (*b*) If I borrow $8500 at 9% interest for 6 months, I will pay $38.25 interest.

The minicalculator can be used as a resource tool that promotes student independence in problem solving.

1. Using the measurements of at least ten different circles, divide the circumference by the diameter in each case and record the quotient. In this way the value of π is clearly illustrated to students.
2. Compute how large a square would be needed to allow every person in the United States 1 square foot to stand on. Compare this to the square that is equal to the area of your state.
3. Work the Lucky Seven puzzle. Push only the 4, ⊞, ⊠, ⊟, ⊡, and ⊜, until you get a 7 on the display. What is the least number of keys needed to accomplish this result?

The minicalculator can be used to solve problems that previously have been too time-consuming or impractical to be done with paper and pencil.

1. Verify the value of e by computing its value in a series approximation.

$$e = \lim_{n \to \infty} \left(1 + \frac{1}{n}\right)^n$$

$$e = 1 + \frac{1}{1!} + \frac{1}{2!} + \frac{1}{3!} + \frac{1}{4!} + \cdots$$

2. Compare the values of $(2^3)^4$ and 2^{3^4}.
3. Solve this problem: A family of eight eats dinner seated around a circular table each evening and wants to use a different seating arrangement each time. How many years (360 days) would elapse before any one seating arrangement would need to be repeated?

The minicalculator can be used to formulate generalizations from patterns of numbers that are displayed.

1.
$$\begin{aligned} 9 \times 9 &= \_\_\_\_\_\_ \\ 99 \times 99 &= \_\_\_\_\_\_ \\ 999 \times 999 &= \_\_\_\_\_\_ \\ &\vdots \\ 999\,999 \times 999\,999 &= \_\_\_\_\_\_ \end{aligned}$$

2. (*a*) Find the decimal form for $\frac{1}{9}$, $\frac{2}{9}$, and $\frac{3}{9}$. Predict the decimal form for $\frac{4}{9}$ and $\frac{5}{9}$. Check your predictions on the machine.

   (*b*) Find the decimal form for $\frac{32}{99}$, $\frac{57}{99}$, and $\frac{43}{99}$. Predict the decimal form for $\frac{76}{99}$. Check your prediction.
3. Compute the following sums and find the generalizations:

$$1^3 + 2^3 =$$
$$1^3 + 2^3 + 3^3 =$$
$$1^3 + 2^3 + 3^3 + 4^3 =$$
$$1^3 + 2^3 + 3^3 + 4^3 + 5^3 =$$

The minicalculator can be used to decrease the time needed to solve difficult computations.

1. What is 11! if $3! = 1 \times 2 \times 3$ and $4! = 1 \times 2 \times 3 \times 4$?
2. Find $\sqrt[3]{7}$ by successive approximations.
3. Find what percentage of the area of a square is covered by an inscribed circle. Is this relation constant regardless of the size of the square? What percentage of the volume of a cube is enclosed in a sphere that just fits in the cube?

A modest proposal concerning the use of hand calculators in schools

EDWIN E. HOPKINS

A teacher of mathematics at Edmondson High School in the Baltimore Public Schools, Edwin Hopkins brings to the teaching of mathematics a background in philosophy. He has taught both mathematics and philosophy on the college level.

Educators and laymen alike have shown a great resistance to the use of hand calculators in schools, particularly in the lower grades. Educators have been willing to admit—or to consider the admittance of—hand calculators only in a case where the calculator adds a "new dimension" of learning to the experience of the child; only in a case where a calculator allows a child to do something, or learn something, that he could not before. I submit that this resistance to the use of hand calculators is irrational and I propose that we make fullest possible use of calculators in all grades of our schools.

I am aware that my position may seem to be heretical and that it might be rejected without examination by many people. The reaction against calculators has been a blind one. The very idea of using calculators extensively in our schools is shocking to many educators. It is too radical a departure from what has been done and what has been accepted concerning our educational system. But we must examine this morbid aversion to calculators. I think that there is nothing to it but fluff and nonsense. (If you are inclined to stop reading at this point because you know that what I have proposed is wrong—if not immoral—ask yourself how you know.)

The most common argument against the use of calculators in the schools is that their use would lead to a decay of the understanding of arithmetic in those students who already knew the rudiments and would prevent the learning of arithmetic in those students who did not. Let us take a look at what would happen if my radical proposal were carried out. Assume that we begin introducing calculators into the curriculum in the first grade (or kindergarten, if you wish). The students would still have to learn the basic facts of addition, subtraction, multiplication, and division. (This really cannot be avoided. After all, the students will be doing algebra one day.) But after learning these facts the calculator would come into play. All addition, subtraction, multiplication, and division of multi-digit numbers would be done on the calculator. Students would not learn, or at least would not become proficient at, the use of the algorithms for multiplication and long division. The children would still have to learn the meanings of the various operations. They would still be working word problems and would still have to learn what words meant add, which ones meant multiply, and so on. This dimension in their learning is present now and would still be present. As their education progressed, they could be given more sophisticated calculators. Let this suffice as a rough outline of what would happen.

If the foregoing situation came into being, and if a child lost his calculator or if the calculator battery ran down, he would not be able to do the basic computations. The student of today is not like this. He can compute without a calculator. This looks like an enormous difference and a great loss, but is it? If you take the student of today and take away his pencil and paper—or if he forgets to bring his pencil to class—he cannot calculate. (There are some students who can do many computations in their head, but these are too few to be significant.) Where is the great difference between the present student and the future one? One needs one sort of instrument to calculate, the pencil and paper; the other needs another sort of instrument, the hand calculator. Neither is able to do calculations without his instrument.

I do not think that anyone would want to attack the calculator on the grounds that it is an inferior instrument to pencil and paper. It must be granted that it is a better instrument. It is faster, more accurate, and in the long run, cheaper. Not only is the price of calculators steadily coming down, but think of all the paper and pencils a student could save if he used a calculator for the grades one through twelve. Calculators could be made tougher, simpler, and ever cheaper. (They also could come with a magnet on the back so they could be slapped inside the child's lunch box.) They would be less likely to be lost than pencils—though this is not making a very great claim.

The aversion of many people to the use of calculators in schools is simply the resistance to a change in *instruments* in calculation brought about by advancing technology. Technological changes are hard to accept at first and one can easily muster up all sorts of specters of doom when first presented with them. Look at the initial reaction of the public—or of much of it—to the horseless carriage. Look at the initial reaction of the public to the use of computers in business and industry. Technological conservatism is not a new phenomenon.

Many would feel that I have not gotten into the heart of the controversy yet. Let me enter in, then. The real problem involved in the use of this new instrument of calculation—and here the reactionaries are speaking—is that the use of the proposed instrument of calculation does not carry with it an *understanding* of the basic operations of arithmetic as does the use of the present instruments—pencil and paper.

My first point—and this is a very important one—is that if you feel that I have now raised the crucial point, ask yourself if you could explain in any detail *at all* what dimensions of understanding are involved in doing arithmetic computations with pencil and paper. Don't say there must be such dimensions of understanding. Have you analyzed the matter and seen what they are? Could you now say what they are? If these questions make you stop and think, then you have not yet the grounds for saying that certain dimensions of understanding will be lost with the advent of the hand calculator. What then is your basis for resistance? Even if you could find such dimensions of understanding now that I am pushing you for them, is it not strange that you resisted the calculator before you knew about them?

Let us push on. Looking at the present day student, there is a difference between his merely being able to calculate and his having some understanding of what he is doing. I grant that there is such a thing as an added dimension of understanding that can sometimes be present in calculating. For instance, the student with a good understanding of place value and regrouping has a better understanding of addition and subtraction than the student who merely borrows and carries. Also, the student who, in multiplying fractions, can explain cancelling in terms of multiplying by 1 has a better understanding of what he is doing than one who can't.

But the foregoing clearly shows that our concern is misplaced when we focus on the instrument of calculation. In using pencil and paper—the present and somewhat sacrosanct instrument of calculation—it is possible to learn the computational skills in a merely mechanical way. Mathematics teachers and others are keenly aware of this possibility and of the fact that this is all that most students possess when it comes to calculation—the mere mechanical skills, with no idea of what lies behind them. (If you have inflated ideas of the understanding of your students of computation, show them

$$\frac{1\not{6}}{\not{6}4} = \frac{1}{4}$$

and see how many of them are grateful to have extended their knowledge.)

Great efforts have been made in recent years to teach mathematics in such a way so that students understand the motions that they go through. These efforts are praiseworthy—it is desirable that students understand what they are doing. To change the students' instruments of calculation is not in any way to undermine these efforts. The efforts would and should still continue, but they would then be focused on getting children to understand what they are doing when they use the calculator, and I suspect that we would be generally unsuccessful, as we are now. The calculator, like the pencil and paper, is merely an instrument of calculation that can be used with little understanding of what is going on. One does not have to be against the calculator to be in favor of increased understanding. These matters are not connected. There always has been the danger that students will learn how to compute without knowing what they are doing, and the use of calculators in schools, while not alleviating this danger, will not increase it.

I am not claiming that nothing will be lost in the proposed transition to hand calculators. Something will be lost. Once we free ourselves of the phobias surrounding hand calculators, we can put the loss in perspective and see how little it will be—and, perhaps, very much is to be gained. A comparison from history is suggestive here. In the *Phaedrus,* Plato has a short discussion on the value of reading, which was at that time just coming into vogue. The tradition before then was oral—one memorized stories and recited them. One of the participants in the dialogue expresses the fear that the coming of written materials will lead to the decay of the ability to memorize and recite works. He was quite right. The ability to do this has generally been lost. But look at what has been gained. The amount of literature that one can absorb has been increased 10 000 fold. I think it will prove to be the same in calculating. Children will lose the ability to do sums and products on paper. However, at a very early age they will gain the ability—and this could be a very exhilarating experience—of doing problems with very large numbers, very quickly and very accurately. Also think of all the time that is now spent on drill in the simple arithmetic operations. This time could be put to good use doing more complicated word problems and perhaps in making a thorough study of the theory of arithmetic. But these suggestions only scratch the surface. After hand calculators have been introduced into the schools in all grades, I am sure that teachers will find a myriad of things to do with them and with the time that is released by the use of calculators. We should not try to prevent the introduction of hand calculators into schools, or fear it, we should, I think, accept it as inevitable and begin a study of ways to make it as fruitful as possible.

Electronic calculators in the classroom

LOWELL STULTZ

Having spent six years teaching junior and senior high school mathematics and six years as a college instructor, Lowell Stultz has most recently participated in an experimental tutoring program using electronic calculators in fifth-grade mathematics classes.

Anyone who has noticed recent newspaper advertisements or attended book and equipment exhibits at teachers' conferences is aware of the exploding market in electronic calculators. The microcircuit technology spin-off from the space industry has given the consumer a handy mathematical tool that will soon become commonplace in the home, office, shop, and classroom, but it is too early to predict the full extent to which the availability of electronic calculators will affect the teaching of mathematics. This article presents some of the uses of calculators in the elementary classroom as well as some of the changes that may become necessary in the subject content itself.

All an educator need do to see some of these new electronic marvels is consult the yellow pages of a city telephone directory. A few calls will result in many salesmen coming with models of calculators for any budget and almost any calculating purpose.

At present, there are several types of calculators ranging in price from \$50 to \$4,000. Optional features—the most notable being the choice of either electronic display or paper printout—are available for each type.

The type most likely to be used in elementary schools is the small hand-sized simple calculator (\$50 to \$150) that performs the four basic operations: +, −, ×, and ÷. Recommended features on these are keys for both "CLEAR" and "CLEAR ENTRY," a "constant" switch for multiplication and division by a constant, and full "floating point" display. Many models of this size also come with rechargeable batteries. Other, more expensive models have a few more features for more complex calculation and would generally be more suited for high school and college.

Most of the calculators suitable for elementary school mathematics have the following four operation keys: (±) (=) (×) (÷). To solve the problem

$$7 - 5 = ?$$

the buttons (7) (±) (5) (=) are pushed. If we use a rectangle for the displayed numbers, the calculation goes like this:

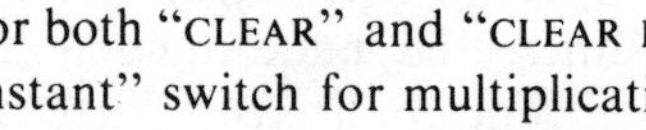

It may be desirable to call attention to the fact that the + and − symbols on the (±) key and the (=) key have a dual role—as the signs of numbers and also as the operations of addition and subtraction.

NOTE: The examples in this article are based on the author's use of a pocket calculator from Sears Roebuck Co., which is identical to one made by Bowmar.

The multiplication of signed numbers for the problem

$$3 \times 2 = ?$$

can be written

$$3^{-} \times 2^{+} = ?$$

and the calculation on the calculator is done in the following order:

Division is identical to the multiplication process except that the ÷ key is used instead of the × key.

Uses in the elementary classroom

Teachers are just beginning to imagine the uses that these devices might have in the classroom. As years go by, we shall see many innovative projects and experiments in which students use calculators. The following list of ideas is a modest attempt to stir the imagination by showing how a calculator may be used as an instructional tool in the mathematics classroom. We shall assume that at least one calculator is available for every two students. Any higher ratio of students to calculator has been found to lead to frustration, confusion, and a dislike for the machine.

1. Preschool, kindergarten, and first-grade students enjoy using the calculator to count. First graders can make up their own addition and subtraction problems with one-digit numbers. Allow the students to work their own problems on paper first and then to check them on the calculator. You can even explain negative differences as "in the hole" numbers, or "owed" numbers.

2. For any grade, the possibility of a student's making up his own problems presents a motivating experience never practical before. The teacher will need to limit the size and order of the numbers and specify the operation. Some students at first will tend to make up oversimplified problems, but this becomes boring after a while. Others will experiment with numbers that are too large or with more difficult chains of operations. Nevertheless, this whets their appetite for more advanced mathematics.

If students are required to do several problems on paper before checking them on the calculator, they will be challenged to work for more speed and accuracy. They often cannot wait to see how well they do with their inventions. Students who get their own problem wrong feel that the right answer on the calculator is something like a "slap on the hand."

3. At any grade level, the calculator may be used not only to check answers but also to debug the problem. In a multiplication problem, a student can check partial products. In addition problems, he can get partial sums as he adds a list of numbers. In division, he can check the products and the subtractions. In larger division problems in later grades, he can use the calculator to help guess partial quotients. Allow the student to do long division using only the × key. He can then use the ÷ to check the answer when finished.

4. The concept of place value is important in all the operations at all levels. The calculator can be used to check the expanded form of a number by adding digits with the appropriate number of zeros after them. For example:

$$5000 + 300 + 70 + 4 = 5374$$

In multiplication, the partial products will add up to the answer only if their place value is preserved. In the vertical form of the problem

$$\begin{array}{r} 354 \\ \times\ 26 \\ \hline 2124 \\ 708\ \ \\ \hline 9204 \end{array}$$

it is easily seen by students that the 708 must be in the tens column. On the calculator, these partial products will add up to the same answer as when the × key is used only if zeros are entered to preserve place value:

$$\begin{aligned} 26 \times 354 &= (20 + 6) \times 354 \\ &= (20 \times 354) + (6 \times 354) = 7080 + 2124 \end{aligned}$$

Note: One difficulty on all small calculators available at the writing of this article is a lack of the capacity to perform the distributive principle for multiplication over addition. For instance, in

$$7 \times 36 = 7 \times (30 + 6) = (7 \times 30) + (7 \times 6),$$

the first two expressions can be performed in a chain of calculations, but the third cannot. There is no way to "hold" one product while another product is obtained and then add them. (This is possible on more expensive machines that have the capability of permanent storage of a constant.) Students can, however, jot down their partial solutions to problems on paper and continue with the electronic calculations.

5. Calculators can serve the same purpose as flash cards with quick oral or written response and immediate reinforcement. Singly, or in a group, students can be asked to enter a number, an operation, and another number. Then after giving the answer, they depress the = key and find out whether they were right.

6. The fact that calculators can display fractions only as decimals is a drawback in long decimals or when division remainders are to be expressed in fractional form. But this feature is also a great aid in introducing decimal fractions in later grades. Very soon a student using a calculator for some division problems wonders about the numbers after the decimal point. Even in early grades, a student can be shown that

$$11 \div 8 = 1, r\,3 = 1 + \frac{3}{8} = 1\frac{3}{8} = 1.375.$$

In the lower grades where a student needs to check a division problem that has a remainder, he can learn to divide the remainder by the divisor using the calculator to see if the decimal fraction is the same as when he uses the ÷ key in the original problem.

7. Most small calculators will do a "chain" of calculations. In the later grades, much can be done with the order of operations on signed numbers using the calculator. If this were a major effort in the seventh and eighth grades, the student's readiness for algebra should be enhanced tremendously. For example, the problem

$$3 - 6\{5 - [4 + 2(7)]\}$$

must be performed in mathematically correct order on the calculator if the student is to be consistent in what he thinks is correct on paper. The order of calculations would be

8. The constant key is one of the more useful features on these small instruments. This key automatically stores the first number in multiplication or the second number in a division problem so that it can be used over again in the next calculation. For instance, after the calculation

pushing 5 = results in [redacted]; pushing only = would multiply the last displayed number by 7. Because it allows for repeated multiplication, this procedure can be used to demonstrate powers of a number, as follows:

When this process is reversed, something very interesting happens: if the ÷ key is pushed, and then the = key, we get next

Continuing, we get still more repeated divisions by 3,

which are approximately the decimal fractions for $1/3^2$, $1/3^3$, and $1/3^4$. If the operation is reversed again by depressing first the × key and then the = key, we get

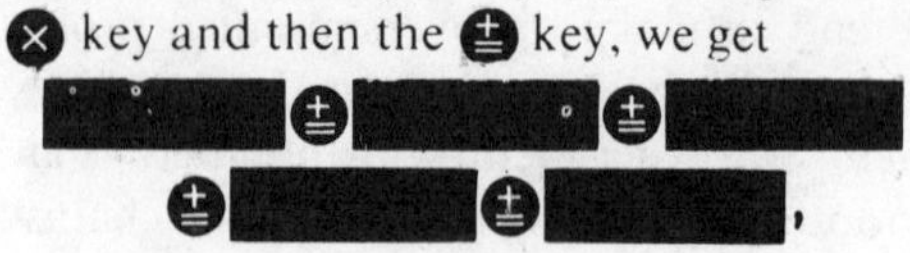

which is not exactly the 3 we started with. With students, this fascinating little experiment leads naturally to the question Why? Now the teacher has a perfect opportunity to talk about number approximations, truncation errors, rounding numbers off, and errors due to repeated calculation with approximate numbers.

In the elementary classroom we have no obligation to go into the higher concepts of "degree of error," "precision," or "range of accuracy," but we shall have to pay the price for the use of electronic machines by discussing their limitations.

9. One of the more practical uses of a calculator in junior high school mathematics is in the evaluation of formulas. Formulas for measurement are encountered in science classes and geometry lessons even though the student has not yet had algebra.

10. "Ah ha! A function machine!" This is what a fifth grader told me when we started working with a small calculator a few days after we had struggled with the concept of a function.

An electronic calculator is a perfect example of the "machine," or "rule," idea of a function. Most elementary textbooks use this concept anyway in pictures and diagrams. The student sees quickly that given a certain input, the calculator can display only one number—it has no choice. If it is to work properly, it has to display something, and the output will always be the same for the same input.

It is worthwhile at this point to ask the student to find something that the machine cannot do. Even the youngest students sooner or later discover that dividing by zero makes the machine display an error code, and in some calculators everything is unworkable until it is cleared. On larger machines with a $\sqrt{}$ button, the square root of a negative number results in an error indication. These experiences lead naturally to a discussion of the domain of the division and square-root functions.

The first uses of functions are usually the basic +, −, ×, and ÷ operations with whole numbers. Here again the calculator can be used to check a student's work and allow him to make up his own domain or range values. For example, define a function $f(n) = 7 \times n$ and ask the student to make up his own partial table like this:

n	$f(n)$
1	_
3	_
0	_
_	49
12	_
_	14
_	56
_	0
11	_

Allow him to fill in the whole table by hand first and then have him check it with the calculator.

11. The graphing of functions also becomes more exciting with the use of a calculator. After the usual discussion of coordinates and the graphing of ordered pairs of whole numbers, ask the student to put random large and small numbers in the calculator function and see if the ordered pair results in a point on the same graph.

With a little thought and ingenuity, the reader can come up with many more applications of these little electronic marvels. The suggestions made in this article would not require a classroom to have calculators all day, every day. A set of a dozen calculators, delivered occasionally to the room like other audiovisual equipment, would be enough to make them available to the whole school.

Why not have a calculator tournament?

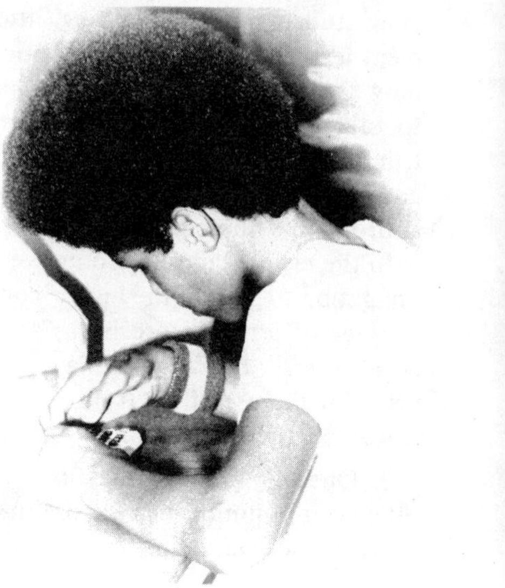

LARRY F. GUTHRIE and
CLYDE A. WILES

An assistant professor of education at Indiana University Northwest in Gary, Larry Guthrie teaches advanced science methods and supervises graduate research in elementary education. Clyde Wiles, also an assistant professor of education and at the same institution, is team leader of the graduate elementary professional sequence. Both were involved with the described calculator tournament.

Why not have a calculator tournament? We did and in a matter of a few weeks had 128 participants from 28 schools registered for the Saturday morning competition. Registration began at 9:00, but contestants and their school sponsors began arriving well before 8:30. The contestants compared notes, calculators, individual preparations, and generally sized up the competition. The calculators they brought varied from the simple $10 variety with 8 digits to sophisticated programmable calculators. The competition started at 9:30 with students being asked to solve as many of the problems as they could in the 45 minutes allotted. Their score was simply the number correctly solved.

During the interim between the competition and an awards luncheon, while the committee was frantically scoring the results, participants and sponsors were involved in a variety of activities. These included movies; how-to-solve-it sessions, which reviewed the tests to provide insights to efficient solutions; and demonstrations set up by university departments. These demonstrations included such things as X-ray defraction analysis, computer games, and the mathematics of population change. Forty-five awards were presented in three categories at the noon luncheon. By 1:00 p.m. every third contestant was proudly displaying a trophy or ribbon. Sponsors were beginning to talk about preparing for next year's competition and the tournament committee was seen sitting back looking quite exhausted and pleased.

Our tournament was sponsored by the Gary Area Council of Teachers of Mathematics and was hosted on the campus of Indiana University Northwest, Gary, Indiana. Committee members who planned the tournament included university personnel and secondary and elementary mathematics teachers. Each school in the local area was invited to enter three participants and one alternate in each appropriate category. A fee of $1 per participant was charged to cover the cost of awards with the university providing the free luncheon to support the effort.

The three categories for the competition were as follows:

The open category I. This category was for grades 5 through 12. It allowed the use of any battery-operated calculator and contained the most complex problems.

The restricted category II. This category was for grades 9 through 12. It allowed the use of only four-function calculators and contained appropriately restricted problems.

The restricted category III. This category was for grades 5 through 8. It also allowed only a four-function calculator and contained problems appropriate for the grade levels.

The minicalculator has invaded society at large and the classroom in particular. And while it may be viewed with dismay by many, it is a modern reality that must be reckoned with. Recent reports indicate that it can be a creative tool for classroom use and instruction. See for example the articles by Machlowitz (1976) in the *Mathematics Teacher* and Stultz (1975) in the *Arithmetic Teacher.* Rising interest and reduced costs have made the question, Shall we use minicalculators? empty. The only practical question is, How shall these machines and their even more sophisticated relatives (Johnsonbaugh 1976) be integrated into the mathematics curriculum in appropriate ways?

It is not our purpose to debate the issues, but we will take a position.

1. The minicalculator should be used to remove the tedium of long or involved sets of calculations. This use, however, should not become an excuse to avoid learning the concepts, processes, and patterns of computations.

2. Instruction involving the calculator should emphasize its power to perform some tasks quite nicely and to make feasible applications where insight and progress may be severely hindered by computational complexity. Simply said, the calculator must become a usable tool and the student a craftsman who understands both its possibilities and its limitations.

3. The calculator should become a source of new mathematics; it should be used to pose and solve problems that are beyond its immediate capacity.

What then do you do with a calculator tournament? While some may suggest a simple test of "finger facility" where 40 calculations per minute replace 40 words per minute, this seems to be a far too narrow view of the possibilities. To expand this view, categories for competition can be identified along several dimensions. For example, we used the achievement background of the student, the complexity of the calculator, and the type of problem in the following ways:

Achievement level. For our purposes, grade level was used as an indicator of achievement. The categories 5–12 open, 9–12 restricted, and 5–8 restricted were found to be too broad. We suggest that categories be more narrowly defined, such as 5–6, 7–8, 9–10, and 11–12.

Complexity of the calculator. Since the prices of even complex machines are dropping rapidly, complexity is an important variable. It is critical to consider the number of digits displayed, the variety of functions, memories, and so on, in order to prevent some participants from having undue advantage. We limited the restricted categories to only four functions. In any event, it is suggested that categories of calculators be matched appropriately to the type and complexity of the problems.

Type of problem. We identified three types of problems: those involving only basic computation or button-pushing skills, those requiring the application of mathematics knowledge to more complex problems, and those that require the student to go beyond the capacity of his machine. Samples of each type are included in table 1.

The three dimensions just identified can be combined to provide many possible categories, some of which may not be useful. Local objectives may suggest combinations or dimensions different from those we chose. For maximum success, interest, and motivation, categories should provide competition for a broad range of participants.

Table 1
Sample problems

<table>
<tr><th></th><th colspan="3">Category</th></tr>
<tr><th>Problem Types</th><th>I</th><th>II</th><th>III</th></tr>
<tr><td>Computation</td><td>Every row, column, and diagonal of a magic square has the same sum. For this square the sum is 15.
<table><tr><td>2</td><td>7</td><td>6</td></tr><tr><td>9</td><td>5</td><td>1</td></tr><tr><td>4</td><td>3</td><td>8</td></tr></table>Find the number that must be placed in the ○ to make this a magic square.
<table><tr><td>154</td><td>77</td><td></td></tr><tr><td>99</td><td></td><td>○</td></tr><tr><td>110</td><td></td><td>88</td></tr></table></td><td>Find the value of
$(6^2 \times 45) - (165 \times 5) + \frac{2^4 \times 5}{8}$</td><td>$33 - (17 \times .907) = ?$</td></tr>
<tr><td>Application</td><td>One thousand meters of cable are to be rolled onto a 1-meter spool that has a radius of 30 centimeters. The cable is 1 centimeter thick. To the nearest tenth, how many layers will there be?
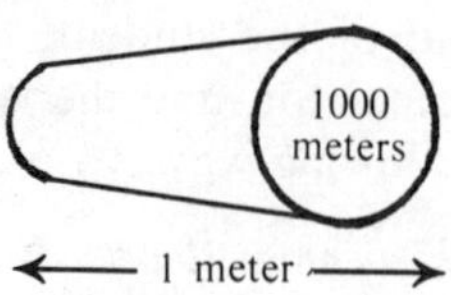</td><td>A two-lane traffic jam of $3\frac{1}{2}$ miles length is observed on the interstate. If we suppose an average distance of 8 feet between cars and an average car length of $12\frac{1}{2}$ feet, how many cars are involved in this jam?
(1 mile contains 5280 feet)</td><td>In this problem there are three missing digits. Find the missing digit in the number
□ 64
2 2 6
4 3 □
1 2 8
4 8 6
3 □ 9
□ 6 4
+ 5 2 2
2 9 4 8</td></tr>
<tr><td>Extension</td><td>Find 11^8</td><td>If $1! = 1$, $2! = 1 \times 2 = 2$, $3! = 1 \times 2 \times 3 = 6$, $4! = 1 \times 2 \times 3 \times 4 = 24$, and so on, find 17!</td><td>Find the decimal expression for the number n correct to the nearest hundredth (two decimal places).
$n \times n \times n = 5$</td></tr>
</table>

PLANNING A TOURNAMENT

Suppose you decide to have a tournament. What must you do? We found six matters required attention. You may need to make adjustments depending on the scope of the tournament you plan.

Committee

You have to have a committee of motivated, hard-working people. The committee should be large enough to cover the entire range of schools and levels to be involved. And, of course, you must have a chairman to oversee the tournament, assign tasks, and encourage committee members. It's also nice to have someone around designated chairman to handle all the problems you don't want to deal with at the moment. Organize early; three months advance planning is not too much.

Publicity

Contact all potential participants. By sending information and registration forms to each mathematics teacher and each building principal in the area we had almost total coverage of the student population. Lists of mathematics teachers and principals were made available from the State Department of Public Instruction. In addition, we found newspapers and radio stations eager to pick up coverage of the event.

Facilities and program

Choose comfortable rooms for the competition that provide adequate working space and access to a clock. Provide proctors and minimize disruptions. We found that one room for each category worked very well.

Don't leave program gaps. The interim between the competition and presentation of awards provides an excellent opportunity to involve both students and school sponsors in a variety of meaningful activities. The type and variety will depend on your local resources and ingenuity. The potential of this period of time should not be overlooked.

Registration

Each school was invited to enter up to three participants and one alternate in each category. Registration required advance registration forms prior to the tournament so that name tags could be prepared. The name tags were coded by color to indicate the category of the contestant and served as a lunch ticket. During registration, each participant filled out a brief form including name, school, and category to validate his or her attendance.

Participants selected among the program options by choosing color-coded tickets as they validated their preregistration. Limited numbers of tickets were prepared for each option. This procedure increased the commitment of the participants to attend the sessions and equalized the numbers attending each session.

Registration was completed by typing a validated roster of schools and participants on spirit duplicating masters. The final results were later recorded on these masters for duplication and distribution to sponsors.

Test construction and scoring

We attempted to prepare tests that would insure the success of participants on the initial portion of the test and tax even the best students on the latter portion. No one was expected to solve all the problems. To help generate appropriate problems, each sponsor was asked to submit four sample problems with each preregistration form. A testing period of 45 minutes was allowed. This time period was sufficient to discriminate among competitors without unduly frustrating those who had done all they could in the first 30 minutes.

Accurate keys and scoring are critical for a successful tournament. Keys should be checked by several competent persons. Each paper should be checked at least twice. To handle the task efficiently, it is suggested that scorers be organized into teams for each category. There should be one grader for each 8 to 10 participants.

Results and awards

Engraved trophies and gold embossed ribbons are both attractive and inexpensive. Trophies were presented to the top three individuals in each category and ribbons to places 4 through 12. In addition, awards were presented to the top three schools in each category on the basis of total team score. The total cost of awards was approximately $75.

Scores and ranks for each participant were recorded on the validated roster and duplicated. These duplicated summaries were placed in a sponsor's packet that was distributed during the awards presentation. The packet also included a copy of each test and a participant award for each contestant.

A planning committee of eight persons attending to these six matters made our first annual calculator tournament a success. Other volunteers were involved in the actual implementation as specific needs arose. While a tournament is a lot of work, an organized committee of enthusiastic teachers places the task well within the capabilities of a local group.

The lack of experienced people should not stand in the way of a successful tournament. We observed a tremendous amount of enthusiasm generated as a result of the tournament. The enthusiasm took many forms. For some participants it was simply the competition and recognition involved; for others it was a Saturday morning spent with peers and a concerned sponsor. Teachers and students found the preparation and practice that preceded the tournament to be highly motivating, while many expressed appreciation for the program that followed the competition.

Sponsors also had an opportunity to attend portions of the Indiana Council of Teachers of Mathematics Regional Conference, which was held in conjunction with the tournament. They were able to renew acquaintances and share ideas and concerns in a relaxed and informal setting. Since the tournament was held in a university setting, it served to alert the university community to current concerns of teachers; it also alerted teachers to the immediately available resources of the local institution.

So, why *not* have a calculator tournament? We trust that this report will encourage individuals and groups of teachers to investigate the possibilities of a calculator tournament in your school, city, or region.

References

Johnsonbaugh, Richard. "Applications of Calculators and Computers to Limits." *Mathematics Teacher* 69 (January 1976):60–65.

Machlowitz, Eleanore. "Electronic Calculators—Friend or Foe of Instruction." *Mathematics Teacher* 69 (February 1976):104–6

Stultz, Lowell. "Electronic Calculators in the Classroom." *Arithmetic Teacher* 21 (February 1975): 135–38.

Calculators for Classroom Use?

By **Eli Teitelbaum**

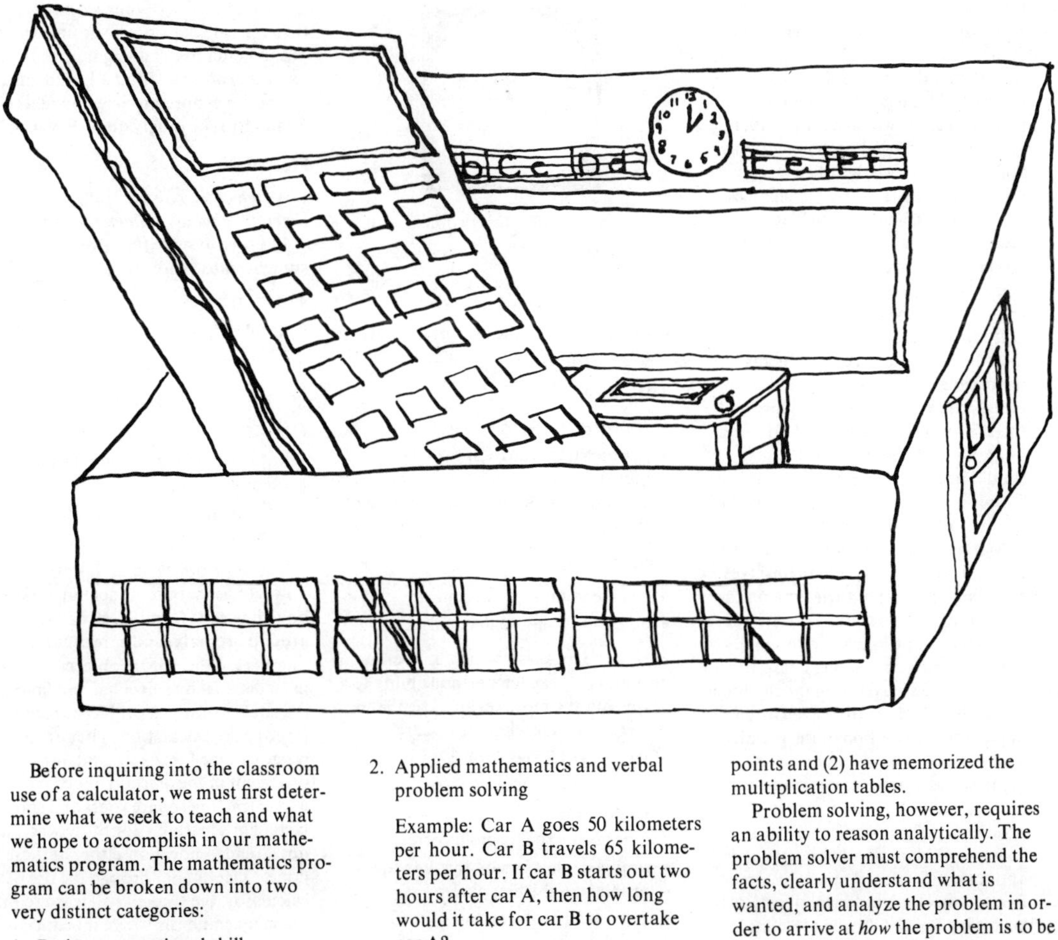

Before inquiring into the classroom use of a calculator, we must first determine what we seek to teach and what we hope to accomplish in our mathematics program. The mathematics program can be broken down into two very distinct categories:

1. Basic computational skills

   Addition, subtraction, multiplication, and division as they relate to whole numbers, fractions, decimals, percentages, and so on.

   Examples:

   $0.7 - 0.01 =$

   $8 \times 1/4 =$

   $4/5 \times 2/6 =$

   $0.01 \overline{)17.2}$

2. Applied mathematics and verbal problem solving

   Example: Car A goes 50 kilometers per hour. Car B travels 65 kilometers per hour. If car B starts out two hours after car A, then how long would it take for car B to overtake car A?

How do these two very distinct areas of the mathematics curriculum relate to the student?

Learning basic computational skills is mostly a matter of memorizing tables and rules. Much of the teacher's work involves drilling students and inducing them to memorize facts. To acquire the mechanical, computational skills, much drill is needed and the student with a good memory has a decided advantage. For example, to divide 0.12 by 1.03, a student must (1) have memorized the rule for moving decimal points and (2) have memorized the multiplication tables.

Problem solving, however, requires an ability to reason analytically. The problem solver must comprehend the facts, clearly understand what is wanted, and analyze the problem in order to arrive at *how* the problem is to be solved.

Students with sharp, analytical minds will do superior work when solving verbal problems, although they may sometimes lack the basic computational skills needed for problem solving. To other students, however, memorization of facts may come quite easily; they soon become proficient in basic computational skills even though they may fail to grasp what should be done in a verbal problem.

Mathematics teachers too often integrate these two completely separate aspects of mathematics, never differen-

Eli Teitelbaum is a mathematics and science instructor at Yeshiva Torah Temimah in Brooklyn, New York.

tiating between the two. Should equal emphasis be placed on the two different categories? Where should the priorities be? Remember! We are teaching mathematics so that it can be used in real-life situations. Therefore a much greater emphasis must be placed on comprehension, analysis, and reasoning than on the mere memorization of rote skills. (This is not to suggest that computational skills are unimportant, merely that the major emphasis in teaching should not be placed on them.) Real life will demand a much greater use of verbal problem-solving ability than knowledge of basic computational processes. After all, chances are slim that calculators will be outlawed at grocery store checkout counters. The logical conclusion is that students would be best served by rigorous training in problem solving.

It is ironic that students who are knowledgeable in verbal problems but make careless mistakes in their computation, should lose points equally with their fellow classmates who hadn't the faintest idea of how the problem should be attacked in the first place. Quite an absurd situation, but a fact of life in most classrooms!

Mathematics applications are hard to teach, yet students find themselves spending far more of their time on the rote (computational) than on the analytical part of problems. In too many cases, the time-consuming computations involved may automatically block the students' minds from even attempting to find the solution of the problem. This is where the calculator has its greatest use. It relieves students of the tedious computational factor in the problem and lets them concentrate on *how* the problem should be attacked and solved. Students' minds become free to think and understand instead of becoming bogged down in tedium.

Mathematics teachers are forever complaining that there is not enough time to spend on verbal problems. With the use of a calculator, the available time can be tripled or quadrupled. Many more problems can be covered in a class period.

Uses of Calculators

There are many ways in which calculators can be used to advantage in teaching mathematics, as even the few examples included here can show.

Problem 1

If the radius of the circle inscribed in a square (fig. 1) *is 2.2 centimeters, what is the area of the shaded region?*

Fig. 1

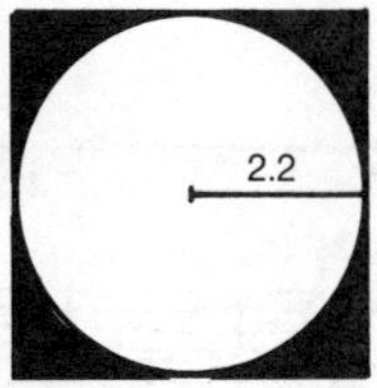

In order to solve this problem, students must know—

- the formula for the area of a circle ($A = \pi r^2$) and how to use it;
- the formula for the area of a square ($A = s^2$) and how it is used.

They must—

- realize that $s = 2r$;
- reason that the area of the circle must be subtracted from the area of the square in order to find the area of the shaded region.

And they must—

- be able to multiply decimals;
- be able to subtract decimals.

If a student gets a wrong answer, it becomes the teacher's responsibility to search out the mistake, analyze it, and correct it. The teacher must search for and allow for a student who—

- may not know how to borrow when subtracting;
- is not using the formula correctly;
- simply cannot reason through the problem in the first place;
- can do the problem, but made a minor slip.

For a teacher with twenty-five students in the class, to look over each student's paper step by step is quite unrealistic. Allowing the student the use of a calculator may not pinpoint a student's difficulty, but it surely narrows the search and reduces the time it takes to locate the difficulty. It has other advantages, too.

(1) Weaker students (those poor in basic computational skills) also feel more confident and they can give the problem proper thought when they have the knowledge that they will not foul things up in the mumbo jumbo of multiplication, division, addition, or subtraction.

(2) Advanced students (proficient in basic computational skills) need not expend a great deal of time working through calculations. Their time can be put to better use in doing more advanced problems. After all, students do not become more proficient in mathematics by repeatingly doing $9 \times 4 = 36$.

Problem 2

The area of a circle is 473.2 square centimeters. Find the diameter of the circle.

In order to solve this problem the students must know—

- $A = \pi r^2$;
- given $A = \pi r^2$, how to turn it around to

$$r = \sqrt{\frac{A}{\pi}};$$

- $d = 2r$;
- how to divide one decimal by another decimal;
- how to find the square root of a number.

The main objective in this problem, however, is that the student understand how to use the formula, switch it around properly, and substitute the numbers in their proper places. Dividing a decimal by a decimal and finding the square root of a problem are matters of pure calculation. They are a waste of time for the student already proficient in the division of decimals and extracting square roots and should be taught separately for those who may not already know how either of these is done. Once again, through the use of a calculator, the student will learn to put the main emphasis where it belongs—on the problem.

Problem 3

Change 11 001$_{two}$ to base ten.

This problem requires—

- knowing how to make a table for powers of two:

2^6	2^5	2^4	2^3	2^2	2^1	2^0

- calculating the powers of two:

64	32	16	8	4	2	1

- understanding place value numeration:

$$(1 \times 16) + (1 \times 8) + (1 \times 1)$$

- multiplying;
- adding.

Of course, here too, the multiplication and addition are quite simple. The crux of the problem is *what* should be multiplied and added. So why not concentrate on that by putting the emphasis there and leaving the simple rote work to the calculator?

Problem 4

Two cars start from point A *and drive to point* B, *a distance of 800 kilometers apart. Car 1 travels at a speed of 60 kilometers per hour and starts out at 9:00 a.m. Car 2 travels at a speed of 80 kilometers per hour and starts at 10:00 a.m. At what time will car 2 overtake car 1? How far will they then be from point* B?

Even though this problem calls for the basic computational skills of multiplication, addition, and subtraction, these are certainly not the main skills that we would like to develop when teaching such a problem. This problem demands that students—

- have a clear and comprehensive understanding of rate × time = distance;
- realize that the distances covered by both cars at the time they meet will be equal;
- know how to apply this formula properly and work out the algebraic equation.

r ×	t =	d
60	z	$60z$
80	$z - 1$	$80(z - 1)$

Certainly no harm can be done if the computation part of the problem is done with a calculator since it bears little on the skills being developed in this problem.

Students often become confused when trying to master too many concepts at a time. The calculator is of immeasurable help in alleviating this source of frustration. It permits the rote mathematics of a problem to be dealt with separately, under less confusing circumstances. And for the better students—why bother them with computation at all, except for occasional review?

Checking for Progress

If the mathematics program is separated into two categories, then it follows that mathematics tests should be separated into two categories. Calculators should *not* be used when you are testing for basic computational skills.

For example:

1. 7.3×2.01
2. $1\,1/2 + 3\,1/4 =$
3. $6 + 5.1 + 3\,1/2 =$
4. $7 \div 0.01 =$
5. $3/2 =$

But they should be used when testing problem solving.

For example:

1. How many kilometers will a car travel in 7 1/2 hours if the car travels at a rate of 95 kilometers per hour?
2. $4231_{\text{five}} = \_\_\_\_\_\__{\text{ten}}$
3. $6^0 + 7^2 + 3^4 =$
4. Find the area of the shaded region if the radius of the circle is 3 centimeters (fig. 2).
5. Find the greatest common factor of 105 and 85.

Fig. 2

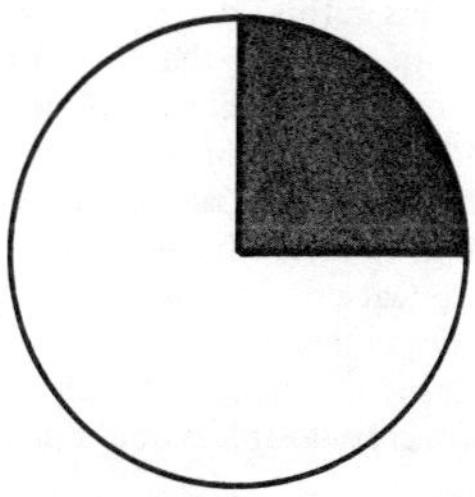

Developing Computational Skills

Many creative educators have written on ways in which calculators can be used to help develop basic computational skills. That subject, however, is not within the scope of this article. Students who are working on their basic computational skills should be encouraged to use a calculator to *check* their answers. Too much class time is spent going over homework and correcting errors which, for the most part, students could have corrected on their own. If an individual student still cannot find his or her mistake, even after analyzing the problem, *then* the teacher can spend the class time in a more meaningful and worthwhile manner by explaining the mistake to that particular student. This saves a great deal of time and allows for more *real* teaching.

Danger of Cheating

This is the greatest fear expressed by many people who regard the calculator as a curse rather than as a blessing. After all, what is to stop students from solving homework problems on their calculators without resorting to any work at all? This is a legitimate point and must be dealt with carefully. Yet, you can carry this question to its logical conclusion and ask, What stops any student from simply copying answers from a friend and not doing the homework on his or her own? It is common knowledge that this is done—even at West Point. Cheating! Of course a student can cheat. But it should be noted that just as a calculator can be used improperly for mathematics homework, in the same manner a student can use a condensed book review for a book report, digest encyclopedia articles instead of researching a report, and resort to a dozen other ingenious tricks to bypass the work of an assignment. Yet in mathematics, more than in any other subject, periodic tests on the basic computational skills (without the use of a calculator) will soon identify those who have been making use of this shortcut. And if these students get 100 percent on their tests, regardless, then more power to them. Either they never needed this kind of practice in the first place or they are being taught below their true level.

This article is very definitely not an exhaustive analysis of all ramifications of the classroom use of a calculator. Suffice it to say, however, that in New York both the Bureau of Mathematics and the Bureau of Curriculum Development of the Board of Education have come out strongly in favor of the use of calculators. That is strong endorsement indeed. As James Fey said, "If you have to ban the calculator to teach a mathematics course, then what you're teaching is trivial."□

Calculators: What Difference Will They Make?

By **Rosemary Schmalz, S. P.**

In any discussion of calculators, we must begin by agreeing that calculators are *in* and that they are never going to be *out*. They may change form, but they will not go away. They are not a fad that will disappear after a certain number of years due to boredom on the part of the American consumer.

Teachers at any level can adopt one of three attitudes:

1. They can consider the calculator a foe and absolutely forbid its use. This, of course, is a losing battle since students can easily use calculators outside school. Such a stand is simply blind.

2. They can admit that calculators do exist and try to use them to advantage, but only as an accessory—a substitute for a log or trig table, a motivator, an enrichment, or a reward.

3. They can integrate calculators into the very body of mathematics instruction at all levels. They can actually develop or adopt a curriculum that uses calculators to teach some concepts.

It is the last alternative that I would like to address in this article.

Teachers must realize that the use of calculators as an integral part of the mathematics instructional program will reorder some basic, age-old traditions in mathematics education. To continue to teach exactly the same mathematics in the same way that it was taught before the advent of the inexpensive calculator just does not make sense.

But what are the age-old traditions that might be changed? Certainly mastery of basic computation facts is not one of them. For a student to need a calculator to rewrite the expression $3(4x + 7)$ would clearly hamper progress in high school mathematics. Without immediate recall of basic facts, students cannot factor or estimate or discover patterns and, in general, students would consider mathematics a chore.

One basic practice that should be considered for change, however, is the choice of algorithm for each of the four operations. What is an algorithm? It is a rule for doing a computation in an efficient and accurate way. The algorithms that we have traditionally used and have passed on to our students are considered to be the most efficient ways to get a correct answer. At least, they take up the least amount of space on the paper. But consider this question: Today, what is the quickest and most efficient way for anyone to get the correct answer to a mulitplication problem having two digits in both factors? Is it not by using a calculator? Therefore, we should take time to look at the algorithms we are teaching. My basic proposal here is that every student should have some algorithm for doing each operation, but it need not be the one that we now teach. Frequently, the rush to get on with teaching the traditional algorithm robs students of any understanding of computation that they have gained in the few lessons leading up to the algorithm. Lack of understanding makes it doubly hard for those students to master the skill. A longer, but more comprehensible, algorithm may be more useful for many students. Let me explain further with a few examples, starting with addition.

An associate professor at Saint Mary-of-the-Woods College in Saint Mary-of-the-Woods, Indiana, Sister Rosemary Schmalz is on sabbatical leave for the academic year 1978–79. As part of her sabbatical program, she is teaching mathematics in elementary schools in Peoria, Illinois.

Most textbooks, after they have introduced a problem involving renaming in the way shown in figure 1*a*, go immediately to the standard algorithm. A few texts show an intermediate step (fig. 1*b*), but never let a student stay with this method for any length of time. Students are forced to proceed almost immediately to the standard algorithm. The method shown in figure 1*b* is an algorithm that will produce a correct answer with little chance of error. It is more directly related to the initial explanations. It is not, however, the most efficient method—in the sense that it takes up more space, consumes an extra line of paper. Yet many students, who could perhaps master this method, are rushed on to the traditional algorithm and never quite master it.

Fig. 1

```
  46       40 + 6
+ 37       30 + 7
          ---------------
           70 + 13
           80 +  3 = 83
               a

  46
+ 37
----
  70
  13
----
  83
   b
```

In subtraction, suppose a teacher convinces his students (perhaps by using examples with a calculator) that $(a - b) = (a + c) - (b + c)$. Then, when faced with a problem like the one shown in figure 2*a*, the student can be taught to add the "magic" number to both numbers, creating a new and easier problem, as shown in figure 2*b*. Of course, the method is less straightforward and takes more paper. For three-digit numbers, it takes two intermediate steps, as shown in figure 2*c*. Yet it is an algorithm that works and many students who have much trouble with the "borrowing" method might be able to master this one.

Fig. 2

```
   82          85
 - 37         -40
 ----        ----
               45
    a           b

  312      314      334
 -178     -180     -200
 ----     ----     ----
                    134
            c
```

Multiplication, also, can be done by a method that is more understandable. For instance, a few texts introduce a problem such as 47×6 by the method shown in figure 3*a*, but they seldom let the child dwell there. Children are rushed on to the familiar algorithm. The same is true for exercises like 47×56. Though the style shown in figure 3*b* may be given as an explanation, it is never intended to be used for more than one lesson or one day. The standard algorithm follows immediately. Yet children who get mixed up on the usual method might do quite well with this method.

Fig. 3

```
   47             47
 × 6              56
 ----           ----
   42             42
  240            240
 ----            350
  282           2000
                ----
    a           2632
                   b
```

Long division is in a class by itself since the standard algorithm is more difficult than any of the others. Yet the idea of division as repeated subtraction is quite simple. If a child can develop skill in doing division by any number of subtractions (as the work is done in figure 4) we should consider him a master of a division algorithm and not push him along to the standard method.

Any discussion of division brings with it the whole question of estimation. Formerly there was an emphasis on estimating in division, but the purpose was misdirected. It was because the efficiency of the algorithm was so important that estimation was introduced at all. Not to guess correctly on the first try was a sign of poor mastery. Yet in the age of calculators, it is important that children can estimate with reasonable accuracy in all four operations. The purpose is twofold. Skill in estimation gives students the ability to check quickly whether the answers they got from their calculators are sensible. More important, it will engender a mind-set in students that will help them to see that frequently there are times when one should not bother to use a calculator to get an exact answer. A good estimate adequately serves the purpose. Such situations appear so often in real-world mathematics that it should not be hard to adopt this as part of the standard curriculum.

Fig. 4

```
        233
 21 ) 4898
      2100     100
      ----
      2798
      2100     100
      ----
       698
       630      30
      ----
        68
        42       2
      ----
        26
        21       1
      ----
         5
```

In summary, every child should develop skill in estimating and should have algorithms that he or she can use in the absense of a calculator. But the calculator must be acknowledged as the most efficient and the usual method of computation. Any difficult standard algorithm should be questioned; it may be replaced by a simpler, easier-to-follow algorithm, even though the latter may take more time and paper. Perhaps the standard algorithms would be shown as points of historical interest in junior high, and bright students could be encouraged to adopt them.

Major changes in such ingrained traditions as the teaching of standard algorithms will not be readily accepted, and if accepted, these changes will radically modify our textbooks and our instruction. But the inexpensive calculator is forcing a revolution and the worst response we can make to it is to ignore it. The most positive response is to look at everything we are teaching in the light of the availability of the calculator and the best interest of our students' futures. □

CALCULATORS IN SECONDARY SCHOOL MATHEMATICS

By MAX S. BELL
The University of Chicago
Chicago, IL 60637

About fifteen years ago R. W. Hamming directed our attention to the notion that "an order of magnitude" change (roughly a factor of ten) in a technology commonly produces fundamentally new effects. As an illustration he offered the tenfold increase in travel speed afforded by automobiles over horse and wagon and the profound consequences of that in our society. He then turned to the "intellectual implications" of computer technology:

> Computers have improved in speed by at least six orders of magnitude over hand calculations—a million fold. In order to understand the factor of a million consider the following two situations: first, you have only one dollar, and second, you have one million dollars. . . . Along with the change in speed there has been a great increase in reliability of operation. . . . Finally, there has been a corresponding decrease in the cost per operation—something more than one thousand times cheaper. It is as if suddenly automobiles now cost two to four dollars, houses twenty or sixty dollars. . . . Our first approach was to carry on in the same old ways, only bigger, faster, and cheaper, and we ignored the order of magnitude change in the new effects. Now we are beginning to explore them. [Hamming 1963]

In the years since 1963 there have been additional order-of-magnitude improvements in computer technology, and computers have indeed had revolutionary effects in very many fields of knowledge and in our society more generally. (It can, of course, be argued that these are not wholly positive effects. See, for example, Weisenbaum [1976].) Even mathematical "proof" has been affected (Appel and Haken 1977). But computers have not yet changed school teaching of mathematics in any fundamental way, despite much interesting work using computers with a relatively small proportion of school youngsters. (See Frederick H. Bell's article [no relation] in this issue.) Likewise, computers as such have not yet had much effect on the everyday calculating activities of ordinary people. Hence, in spite of the computer revolution it has been appropriate for schools to teach personal calculation in much the same old ways, relying primarily on paper-and-pencil algorithms, with occasional assistance from mechanical calculators, slide rules, or books of tables.

Suddenly another technological innovation—hand-held electronic calculators—has substantially altered personal calculation. Furthermore, these calculators are direct alternatives to the arithmetic and calculating methods that up to now have been the principal component of eight years or more of schooling in mathematics. I believe it would be difficult to overemphasize the challenge of that to mathematics education as we now know it. This direct challenge to our profession has come upon us over an amazingly short time period, as indicated by this brief history of the development of calculators:

> It became possible by about 1970 to put all of a basic calculator's logic on a single . . . integrated circuit chip. . . . Within a year that technology gave rise to a new generation of small four function calculators. . . . The prices (well under $200) gave the devices a handhold at the upper end of the mass market. Now that the industry is 100 million units further along . . . the low end of the price range is well below $20. The result is that for millions of people everday arithmetic will never be the same again. [McWhorter 1978]

With the availability of calculators by 1980 almost certain to approximate the availability of automobiles, refrigerators, or television sets, "arithmetic will never be the same" not merely for millions but for essentially every American. The hard won calculation skills of many years of schooling may then be seen by many as irrelevant, and if we have a different view of relevance, we will need to defend it. Furthermore, there is every indication that the calculator revolution for many people outside of schools will go beyond mere replacement of paper-and-pencil arithmetic to minicomputer or hand-held computer power or to automatically programmed special purpose routines in hand-held devices. Also, the imminent availability of interactive visual displays controlled by calculator-like devices and easy access to vast amounts of stored information (including video programs)

Calculation skills may seem irrelevant.

through home or office computer terminals is freely predicted by the communications and microelectronics industries. By now we know it is wise to take such predictions very seriously. But even if nothing more were in store than already existing five- to ten-dollar arithmetic machines and ten- to twenty-dollar scientific or statistical calculators, it is clear that a complete change is inevitable in the ways in which people outside schools cope with arithmetic, the various uses of elementary functions, and statistics.

But that is talk about the world outside of schools and does not necessarily say much about the impact of calculators on education. Furthermore, given our recent experience with such proposed revolutions as instructional television, programmed instruction, and computer-assisted instruction, it is surely risky to predict still another revolution.

Still, with calculators so widespread, so close to what we do in a large chunk of schooling, and so intertwined with what the public believes is nearly the entire purpose of mathematics teaching in schools (calculation skills and their applications), it will probably be impossible for schools to ignore them. Certain effects are already visible, and others seem inevitable. I will outline some of those next.

Effects of Calculators

A first effect of calculators on education is that school mathematics will surely be the center of a lively controversy about how much effect calculators *should* have, especially in pre–high school work. This controversy is likely to be heated and to ventilate again concerns that come up repeatedly in such phrases as these: "People are already too dependent on machines." "Hard things shouldn't be too easy." "It rewards sloth and ignorance to give a calculator to someone who hasn't learned to calculate without one." To some, calculators in schools amount to "pampering," "frills," "waste of taxpayers' money," and other such code words for standard moral concerns about schools. There are also concerns about what should remain as "basic skills" in a computer/calculator age—concerns that must be taken seriously. My own current favorite "solution" to the problem of stating truly basic skills is the National Council of Supervisors of Mathematics (NCSM) "Position Paper on Basic Mathematical Skills," reprinted in the February 1978 *Mathematics Teacher.*

A second likely educative effect of calculators is on adults who are *not* in school or college, many of whom continue to learn on their own throughout their lives. But that continued learning often excludes mathematics; many otherwise capable adults freely admit that they fear mathematics and avoid it where possible, even when the avoidance of numerical information is harmful to them. One can already see that calculators are changing that, however, as in this often repeated "good news, bad news" report from my friends: "I just got a calculator and (bad news) I find it is making me sinfully lazy about doing things I used to do in my head or with paper and pencil, but (the good news) I now tackle things with my calculator that I never would have thought of doing before." I believe that millions of adults will gain through calculators a new ease and sense of control with respect to numerical and other mathematical information.

The third, nearly inevitable, effect of calculators is on formal education at the post–high school level. Most college students taking courses where they would once have used a slide rule or book of mathematical tables now have calculators instead. A couple of years ago there was a brief flurry of doubts similar to those now voiced at earlier school levels: Would students become helplessly dependent on calculators? Would it be fair to use calculators on tests? What of those students that could not afford calculators? Virtually all such "issues" have been settled or will soon be settled in favor of unrestricted use of calculators at the college level. In several large state universities, this is also the practice for thousands of students in "remedial" mathematics courses, and I expect that will also become the usual pattern. I expect that at the college and university level the educational impact of calculators will continue to grow with little controversy or restriction.

In their academic courses, the two-year colleges will use the calculator much the same as their four-year cousins. Likewise, in post-high school training for specific occupations (in trade or technical schools or junior colleges) it seems nearly certain to me that with the pragmatism that characterizes them, such enterprises will quickly elect unrestricted use of calculators wherever they are seen as efficient tools—especially as trade-specific calculators are programmed and marketed.

This brings me to the use of calculators in schools, and here too certain uses of calculators are likely and relatively uncontroversial. For example, secondary school mathematics (and science) courses for the college bound will probably follow the lead of the colleges and universities in putting virtually no restraints on calculator use. At the least, calculators will quickly replace slide rules and books of tables in the service of existing curricula. Also, many teachers of advanced mathematics courses are already exploiting calculators in interesting ways to build concepts, and as programmable models become cheaper the possibilities for that will expand. (For example, on a calculator one can almost *see* the shape of a function or a sequence approaching a limit.) Also, as indicated in other articles in this issue, a nontrivial number of high schools have for some years now used computers in their mathematics programs. Most of those will move readily either to programmable calculators (which will soon cost $30 or less) or to minicomputers, and many who were held back from using computers by expense, limited access, and so on, will likewise exploit such devices.

Even though there will probably be few restrictions on calculators in high school college-preparatory work, one must attend to the response of a certain student when I asked her if a programmable calculator would be helpful in her mathematics courses: "Why bother? There just isn't any calculation to be done that I can't do in my head. Now if you were to ask me about my chemistry or physics courses, I'd have something to say." Indeed we all know that our college preparatory courses generally use easy numbers so exclusively that students who get anything but an integer or easy fraction as an answer to a problem are likely to think they did it wrong! Of course that gives a very distorted view of the applications of mathematics and the sorts of "dirty" data common in applications. Perhaps improvement of that state of affairs by making possible more "real problems with real data" can be one of the first contributions of calculators to the improvement of our college preparatory courses. That student's remarks should also remind us that calculators may at last open the doors to some genuine interaction between science and mathematics courses.

> Unrestricted use of calculators in schools can be expected.

So far I've tried to pinpoint those aspects of education where calculators seem sure to have an impact: people continuing to learn to use mathematics without formal courses, the unrestricted use of calculators in many college courses, their use in trade and technical training, and their use by people in school beyond about age 14 who can already be certified competent in paper-and-pencil calculation. That is, those students who can prove they don't really need calculators can use them without restriction. As often happens, them that has, gets, and the rich (in mathematical power) get richer.

The paradox remains that the part of education most akin to the calculator—the calculation curriculum—is now, and will probably remain, the most stubbornly resistant to any substantial impact by calculators. Yet I believe that the appropriate use of calculators can be very helpful in teaching calculation and its applications, whether with elementary school youngsters or with those mathematical losers who don't get into our college preparatory courses. But in this section we are making predictions about where calculators will *easily* be absorbed into school work, and I think that cannot be said of our present arithmetic and "general mathematics" courses. I have discussed the difficulties in getting calculators used in the calculation curriculum elsewhere (Bell 1976, and Usiskin and Bell 1976).

Needed Curriculum

The news media over the past few years have featured countless articles on calculators, usually with prominent attention given to possible effects on school arithmetic and seldom to effects on other parts of schooling. It would serve no purpose to review that coverage here beyond saying that two polar views have received almost equal coverage. They are well expressed in this early summary:

> Critics argue that students, especially at the elementary level, risk becoming so dependent on calculators that they will forget—or fail to learn in the first place—basic computational skills. . . . Proponents . . . contend minicalculators can be a significant force in moving schools away from "answer-oriented" instruction, freeing both teachers and students for concentration on more important underlying concepts. [*Nation's Schools and Colleges* 1974]

But such a good-bad dichotomy is obviously far too simplistic. The questions are not simply whether to use or not to use calculators, but how, with whom, when, in what ways, and in the service of what objectives.

It is pleasing to report that the mathematics teaching profession saw almost immediately the complexity and importance of the issues surrounding calculators. For example, the NCTM has from the beginning strongly supported full development of the educational possibilities, as must be obvious to anyone who has followed their publications for the past few years. In addition to many articles, several information bulletins, and several policy statements, the *Arithmetic Teacher* in November 1976 published a special issue on calculators. The NACOME report (which every school mathematics department should have on hand) refers in a number of places to the possible impact of calculators (see the October 1976 *Mathematics Teacher*). It also includes one recommendation that has generated considerable controversy, but with which I agree:

> The "calculation" curriculum will be most resistant to change.

> Beginning no later than the end of the eighth grade, a calculator should be available for each mathematics student during each mathematics class. Each student should be permitted to use the calculator during all of his or her mathematical work, including tests. [NACOME 1975, p. 138]

Another useful sorting out of issues was put together by Marilyn Suydam at the request of the National Science Foundation (NSF). It includes among other nice things several appendices discussing calculators and the teaching of arithmetic in far more detail than I can summarize here. For that reason and others, I suggest it as a reference that should be in most school mathematics departments.

In 1976 the National Science Foundation (NSF) and the National Institutes of Education (NIE) sponsored a conference to list some of the research and curriculum development needed for a proper response to calculators. The report on that may be the most comprehensive statement so far of those who believe calculators will and should bring about substantial changes in mathematics education; therefore I will include excerpts from it here and also recommend it as a useful reference for school mathematics departments (*Report of the Conference on Needed Research and Development on Hand-Held Calculators in School Mathematics* 1976).

One thing we must make clear to the public at large is that none of us intends to have calculators entirely take over number work. The intelligent use of the calculator itself and of mathematics in general requires that students have a firm grasp of

Now we can study applications with real data.

such things as "number sense," estimation skills, and a thorough understanding of our number system. To that end the conference listed certain things to be expected of every student, and these resemble other such thoughtful statements from our profession:

1. *Ability to multiply and add one-digit numbers nearly automatically.* A student who has not acquired these "reflexes" is nearly crippled, if only because rapid processing of small-number sums and products is so often a feature of rough-and-ready estimation.

2. *Firm understanding of our standard numeration system, including reading and writing arbitrarily large or small whole numbers and decimals.* As part of this, students should be familiar with notation involving positive and negative powers of 10 (though this is not presently expected), because it helps in estimation and because it is sometimes needed in using calculators.

3. *Firm understanding of the various meanings of fraction notation (e.g., parts of a whole, ratios, division) and ability to exploit "equivalence" of fractions.* Even though there will be more emphasis on decimal notation, fraction ideas remain essential.

4. *Experience with algorithms.* Algorithms in mathematics are an important concern in themselves. But the same sort of work and the same algorithms that are now used should not necessarily be retained. [Conference 1976, p. 13]

Of the twenty-two recommendations of the conference, seven deal specifically with curriculum development:

Curriculum Development for the Immediate Future

Recommendation 6. Materials should be developed to exploit the calculator as a teaching tool at every point in the curriculum to test a variety of ideas and possibilities pending emergence of calculator-integrated curriculums.

Recommendation 7. Research and development should be undertaken with respect to algorithmic processes in school mathematics, especially those in arithmetic and algebra.

Recommendation 8. Curriculum materials should be developed for kindergarten through twelfth grade to teach estimation, approximation, significant digits, order of magnitude calculations, and similar ideas.

Recommendation 9. Curriculum materials should be developed that teach problem-solving strategies more effectively and that build pupils' confidence in their ability to solve problems.

Recommendation 10. Curriculum materials should be developed and tested on new topics that are not now in the curriculum but that calculators may make feasible. The possibilities of adapting "advanced" topics to earlier levels should also be explored.

Curriculum Development for the Long-Range Future

Recommendation 11. Some full-scale alternatives to the K–6 elementary school mathematics program should be developed that use calculators wherever appropriate (but not exclusively) and that considerably broaden the range of mathematical ideas absorbed during the elementary school program.

Recommendation 12. Some new courses for secondary school students should be developed, including consumer-industrial and data-oriented statistics courses, as well as alternatives for junior high school work and for those high school students who do not elect standard "college preparatory" mathematics courses. [Conference 1976, pp. 18–19]

The statements summarized in this section represent only a sample of the responses by our profession to the challenges offered by the invention and rapid dissemination of calculators. Such statements must now be translated into actual materials for use in school.

Conclusion

I have outlined here a strongly procalculator position, though not without reservations. The argument proceeds as follows:

1. Calculators represent the kind of order-of-magnitude change in technology that can substantially change society.

2. One aspect of this change in society is that people outside of school may soon use only calculators to accomplish what we now make the principal component of years of schooling.

3. For schools to ignore this challenge poses unacceptable risks to the proper mathematical education of youngsters.

4. In fact, calculators have already had a considerable impact on education in and out of school. That impact is likely to increase, except in the calculation curriculum, which may be very resistant.

Following that I quoted from several reports that outline some of the short-term and long-term things we may need to do as mathematics educators to bring about an orderly and fruitful adjustment to calculators.

I believe we will succeed or fail according to how innovative and thoughtful many individual teachers are in accommodating their classrooms to the new realities of a calculator and computer age. If we fail, a substantial part of mathematics education may eventually go the way of instruction in handwriting, Greek, or Latin. But if we succeed, the advent of calculators and their microelectronic cousins can help us achieve results we have long hoped for but not attained: Most, rather than a few, people leaving school a decade from now could feel a real sense of pleasure and control with respect to using mathematics in their everyday and working lives.

REFERENCES

Appel, K., and W. Haken. "The Solution to the Four-Color-Map Problem." *Scientific American* (October 1977):108–21.

Bell, Max. "Calculators in Elementary Schools? Some Tentative Guidelines and Questions Based on Classroom Experience." *Arithmetic Teacher* 23 (November 1976):502–9.

Hamming, R. W. "Intellectual Implications of the Computer Revolution." *American Mathematical Monthly* 70 (January 1963):4–11.

McWhorter, Eugene W. "The Small Electronic Calculator." *Scientific American* 234 (March 1978):88–98.

"Calculators in the Classroom." *Nation's Schools and Colleges* 1 (December 1974):12–14.

National Advisory Committee on Mathematical Education. *Overview and Analysis of School Mathematics, Grades K–12.* Conference Board of the Mathematical Sciences, Washington, D.C., 1975. (Copies of the report are available from NCTM (members $2.70, nonmembers $3.00).)

National Council of Supervisors of Mathematics Position Paper on Basic Mathematical Skills. *Mathematics Teacher* 71 (February 1978).

Report of the Conference on Needed Research and Development on Hand-Held Calculators in School Mathematics 1976. (ERIC document number ED 139 665)

Usiskin, Zalman, and Max Bell. "Calculators and School Arithmetic: Some Perspectives," appearing in Marilyn Suydam's "Electronic Hand Calculators: The Implications for Pre-College Education." Final Report for NSF Grant No. EPP 75-16157, 1976. (ERIC document number ED 127 205. The main report or the report with appendices must be specified.)

Weisenbaum, Joseph. *Computer Power and Human Reason.* San Francisco: W. H. Freeman & Co., 1976.

HAND-HELD CALCULATORS AND POTENTIAL REDESIGN OF THE SCHOOL MATHEMATICS CURRICULUM

Rethinking the content and teaching of secondary mathematics in the calculator era.

By **HENRY O. POLLAK**

Bell Laboratories
Murray Hill, NJ 07974

IN RECENT years a number of new techniques and devices to help with teaching have come along, enjoyed a brief period of enormous and somewhat unrestrained enthusiasm, and have then been added in a limited way to our arsenal of available pedagogy. These include, to name just a few, programed instruction, films, and modular scheduling.

It is not entirely unfair to recall that some of these were hailed initially as *the* solutions to all pedagogic problems. In the first wave of excitement, proponents cited many glorious things that could be done with these new techniques to replace or to unburden the teacher in one way of another. This enthusiasm is perfectly natural, and I do not wish to belittle it.

It is unfortunate, however, that in many cases mathematicians reacted to these new techniques with either missionary enthusiasm or uncompromising disdain, rather than taking a deeper look at the problems to which these innovations might be addressed. For example, when the wonderful technique of making films (along with adequate funding) became easily available, many mathematicians joyously jumped in. They thought of all sorts of beautiful phenomena they could present on film, and they had a wonderful time. Unfortunately (and it is very much easier to say this in retrospect), they did not choose to ask the pertinent question: *What are the most difficult pedagogic problems we have in mathematics education, and how can films help us overcome them?* If we had stopped to consider the matter from this angle, we might have made films on conditional probability or curvilinear motion, or might even have developed a brief segment of animation to help with that nasty problem of the volume of the intersection of three mutually perpendicular cylinders, rather than making films on area and limits. The latter are perfectly fine topics, but in presenting them I see no clear advantage for film over a blackboard. We might instead have heavily emphasized three-dimensional and motion problems in our films—that's where the blackboard really can't compete.

What are some of the most difficult problems we have in teaching school mathematics with which the calculator might help?

We find ourselves at this time with a great opportunity opened up by another technique, another device to help us with our teaching—namely, the hand-held calculator. Once again it is perfectly natural to discuss all the glorious things that one might be able to do and all the ways in which some roles of the teacher might be changed. However, I should like to urge that we consider the other side of the coin. What are some of the most difficult prob-

This article was originally contributed by the author independently as an appendix to "Electronic Hand Calculators: The Implications for Pre-College Education," a report prepared under NSF Grant No. EPP 75-16157. The opinions expressed herein do not necessarily reflect the position or policy of the National Science Foundation.

lems we have in teaching school mathematics with which the calculator might help? To keep you from reaching the immediate prediction that this might be the empty set, let me give an example. I am told that we often have great problems in teaching the notion of a function. It is difficult for the students to get a clear hold of the idea that what matters (in the simplest case) is that when a given number is fed to the function, a single number comes out. In fact, we always get the same single answer for a given number that we use to start the process. We try to get the students to realize that any way of describing a specific relationship between the number that goes in and the one that comes out is perfectly fair game and leads to the same function. For example, we may describe the function in words or by a table, a graph, a formula, an arrow diagram, or various other techniques.

One more useful device to add to this collection of alternatives will turn out to be the hand-held calculator; a function can be described by a fixed routine that is followed on the hand-held calculator, a determinate series of buttons to push that yields the correct answer. The calculator approach may also help to prevent the misconception that a function is only a formula.

an opportunity . . . to understand inverse functions by actually experiencing the process

This may seem like a relatively small improvement, but I believe strongly that its physically active nature will make this the best approach to the study of functions for some students. However, in the study of inverse functions I can see the possibility of a real improvement for everybody. My impression is that our analytic ways of describing inverse functions have often gotten bogged down in notation and have been difficult for many students. After all, the statement $g(f(x)) = x$ is a little much for the student who is still very uneasy about $f(x)$ itself. Reflecting the graph across the line $y = x$ is also difficult for those who are struggling with variables and what they mean at the same time they are fighting the $f(x)$ notation.

The hand-held calculator provides an opportunity for the student to comprehend inverse functions by actually experiencing the process. The original function starts out with a number and gives us a second number. The inverse function takes the second number and gives back the original. You can, of course, see this immediately in the use of x^2 and $\sqrt{x}$ keys, of the exponential and logarithmic keys, or of the keys for various direct and inverse trigonometric functions. But there are many fancier things you can do. The inverse of

$$y = 8x^3 - 36x^2 + 54x - 27$$

is

$$x = \tfrac{1}{2}(y^{1/3} + 3).$$

The inverse of

$$y = \sin \frac{x}{1 + x}$$

is

$$x = \frac{\text{arc sin } y}{1 - \text{arc sin } y}.$$

My hunch is that if students experience the transformation of numbers by such pairs of programs on the hand-held calculator and see how it all comes out, they will get a good understanding of inverse functions—one that is hard to obtain any other way.

Besides the topics associated with the understanding of functions there are many others that immediately come to mind in which the hand-held calculator might provide real pedagogic advantages. To name a few, we might now be better able to teach iteration methods for solving simultaneous linear equations or, later on, some more general nonlinear equations. In the learning of probability, the hand-held calculator might greatly increase the variety of experiments that can be the source of data to be studied and used as illustrations. In practical statistics, data analytic computations might become more accessible. So might linear programming. This list is not meant to convey any lengthy consideration of the subject. It is simply given to confirm the notion that there are many mathematical topics with real pedagogic difficulties where the hand-held calculator might help.

If further reflection indicates that a rich variety of such opportunities actually exists—as I believe it does—this opens up the possibility to rethink the curriculum in a much broader sense. I believe the curriculum should be based in a fundamental way on two partial orderings, one of which is essentially supplied by the discipline and the other by society.

We might be better able to teach iteration methods for solving simultaneous linear equations.

The partial orderings I have in mind are those of prerequisites and of importance. It is often true that one mathematical topic really has to precede another, and the design of the curriculum must take this into account. It is up to the experts on the mathematical and pedagogic sides of the house to make clear the existence of such prerequisites, although it is worth pointing out that these experts sometimes change their minds. For example, we once assumed that work with fractions must precede any work on probability. When we finally realized that probability is perhaps the best available *motivation* for work with fractions, we began to experiment seriously with the opposite order.

This does not, however, detract from the point: there is among mathematical topics, or clusters of topics, a partial ordering of prerequisites. Within a cluster, different orders may indeed be practical. The second partial ordering, as I have indicated, is one of importance.

There are some mathematical ideas, topics, techniques that are more essential than others for the population as a whole. For example, to take a simple but not altogether untimely example, I would maintain that probability is more important for the population at large than the division of polynomials. Unfortunately, not very long ago, we taught division of polynomials in the ninth grade and did not teach probability until much later, if at all. Division of polynomials is probably needed for the first time for partial fraction expansions in second-year calculus (that's the prerequisite side of the argument) and not very important in its own right; it is likely that probability is needed for the first time in the elementary school and is enormously important for everybody. Of course, these were probably not common opinions when the traditional mathematics curriculum became solidified. But societal needs, and societal views of the mathematical sciences, have a way of changing, and they influence the mathematics curriculum through this second partial ordering of importance.

societal views of the mathematical sciences have a way of changing

If my views on the pedagogic possibilities of the hand-held calculator are realistic, then these little machines can have a major effect on both the partial orderings we have been discussing. The ability to handle numbers and functions and algorithms in new ways will loosen and even alter some of the prerequisites we have always believed necessary. We shall be able to use the ability to work with division of numbers and with trigonometric functions and with exponents and logarithms as motivation for a deeper study of these topics rather than having to insist that the deeper study must precede any practical use of the functions and techniques. Not that we shall necessarily want to do this always. However, the possibility of inverting the order of experiences gives us the flexibility to reconsider much of the curriculum.

Similarly, the hand-held calculator may allow us to take up a number of topics early, topics that are of great importance to various populations but that we have not been able to approach previously. These are, for example, some of the mathematical topics mentioned previously and with which we have had pedagogic difficulty in the past. To mention just one, if some rudimentary ways of looking at data are more important to everybody than factoring, then perhaps with the aid of the hand-held calculator we can do data analysis earlier and more successfully, and demote factoring to a less prominent position.

There is no obvious single way in which the community that plans for education in the mathematical sciences must organize itself to take advantage of the new opportunities deriving from the easy availability of hand-held calculators. Let me mention just one possibility. It might be sensible to have a conference, in the general pattern and scale of the Cambridge conferences during the previous decade, to reconsider the school curriculum with this new point of view in mind. (Of course, we would need a greater proportion of people knowledgeable in the many aspects of mathematics education than the first Cambridge conference included.) A month in some relatively secluded spot, with days and even nights devoted to a churning and bubbling of ideas, is not a bad way to proceed, although admittedly such a conference would not be a trivial thing to organize. Out of this conference could come a blueprint to help guide the pattern of education's evolution in the mathematical sciences for the near future.

Calculator Activities

Calculator activities in this chapter are organized sequentially from elementary school through high school. Judd presents a collection of seven games for elementary pupils with variations for junior high school students. He also includes suggestions for introducing the games to a class. An activity from the "IDEAS" section of the *Arithmetic Teacher* provides an opportunity for children to use the calculator to practice basic skills. Bruni and Silverman contend that the calculator can be a useful tool to help children understand basic algorithms. They also present several games that help develop skills in estimation, mental calculation, and problem solving.

Working with the calculator can be fun, as evidenced by Friesen, who provides a check for the answers to computation problems that involves turning the calculator upside down to read the corresponding word answer. The calculator can be incorporated in the existing curriculum to develop problem-solving skills, states Morris, and she provides some activities to illustrate how this can be accomplished. Hobbs and Burris use repeating decimals as an example of how the calculator can be used effectively to encourage student inquiry and creativity.

An activity for children in the upper elementary grades is presented by Lappan and Winter. The calculator can be used to do arithmetic in bases other than ten, as illustrated by Schultz. The December 1976 *Mathematics Teacher* includes activities by Woodburn that require hand calculators. Ockenga suggests games and activities for the following topics: computation and estimation skills, measurement and geometry, functions, and problem solving and applications.

Ten classroom activities for use with minicalculators are offered by Litwiller and Duncan. The activities include interesting problems that would be extremely difficult without calculators. Another creative use is presented by Eimer, who explains how to find cube roots with a square root key. Waits extends the method used by Eimer and presents a more direct method for approximating the *n*th root of any positive number using only a four-function calculator with a square root key. Five articles taken from the "Activities" section of the *Mathematics Teacher* provide varied calculator activities.

Recreational activities are presented by Boyle. Kahn provides a motivational exercise using the calculator as a calendar. An investigation of the magic squares of Old China is simplified with the calculator, as exhibited by Swetz. Students can investigate many topics in number theory with the aid of the calculator, and Lichtenberg shows one of these by looking at repeating decimals.

Maor suggests that through the calculator we can introduce numerical analysis, as well as some of the more exciting chapters in the history of mathematics, to our classes. According to Keller, the calculator can be a helpful tool for generating tables. Friedlander suggests some efficient ways to use the calculator by looking at algorithmic design. Programmable calculators are useful in creating mathematical proof, as the article by Snover and Spikell illustrates. Hiatt emphasizes that the calculator can be used to develop the method of inquiry used in mathematics. He suggests that finding areas under curves is one activity that accomplishes this. The calculator as a motivating device in the study of the elementary theory of limits is presented by Johnsonbaugh, who also illustrates how the calculator can be integrated with the theory of limits. Morgan and Warnock advise caution in using the calculator to demonstrate simple limits.

Instructional games with calculators

WALLACE JUDD

An author and consultant, Wallace Judd has worked with students in many classrooms over the past two years developing concepts with the hand-held calculator.

This is a brief collection of calculator games that I have played with kids (and adults) over the last two years. Each game can be explained simply enough for first or second graders to understand, yet the games can be played with variations difficult enough to make most mathematics teachers stop and think.

The machine requirements for all the games are minimal. Only the "Before" and the "After" games require a constant for addition and subtraction; all the others require at most a constant for multiplication and division, which is standard on most small calculators. None of the games require a machine memory, and all of them can be played in less than five minutes.

The kids with whom I have played these games have enjoyed them very much. I hope you do, too.

NIM

Teaches: Addition and place value concepts

For: Two players and one calculator

Object of the game: To get **67** on the display

How to play: The first player pushes a single digit key (not zero), then pushes the ⊞ key. The next player takes his turn by pushing a single digit key (again not zero), then pushing the ⊞ key. Players take turns until a player pushes the ⊞ key and the display reads **67**. The player who pushes ⊞ and gets the display to show **67** wins. If a player pushes ⊞ and the display shows a number larger than **67**, that player has gone "bust" and loses.

Variations for primary grades. Use only the first row of digits—the 1, 2, or 3 keys—and **21** as the goal.

Variations for junior high. Use the first column of digits—the 1, 4, or 7 keys—and go to **47**.

WIPEOUT

Teaches: Place value

For: One number-giver and any number of players, each with a calculator

Object of the game: To remove one digit from the display without changing any of the other digits.

How to play: The number-giver picks a number, which all players enter into their calculators, and says which digit is to be removed. Good numbers are those that have a pattern, so people can tell easily if any digit but the selected one has changed.

Example: In the display, **876543**, wipe out the **7** without changing any other digit.

This is done by subtracting one number from the number on the display. So key in ⊟, then the number to be subtracted, and press ⊜. Does the display read **806543**?

People can take turns giving numbers to each other and selecting the digit to be wiped out.

Variations for grades 2–3. Limit the display to three digits.

Variations for junior high. Use decimals in the display.

Example: Wipe the 8 out of **.567891**.

BEFORE

Teaches: Counting and place value concepts

For: Any number of players and a calculator that has a constant for subtraction. (To see if the calculator will work for this game, key in 1 ⊟ ⊜ ⊜ ⊜ ⊜ ⊜. If the answer is **0** and does not change, the calculator will not work. If the answer changes from **0** to **−1** to **−2** to **−3** to **−4**, . . . , then the calculator will work for this game.)

Object of the game: To predict what number will show on the display when ⊜ is pushed

How to play: Start the game by keying in the sequence 1 ⊟ ⊜. Do not push [C], the "clear" key, during the whole game. Check to see if the calculator is set by keying in 10 ⊜. If the display does not read **9**, something is wrong. Clear the machine and start over again. Key in a number and ask the students what number will come when the ⊜ is pushed. See who guesses correctly. Keep score if you want to make a competition out of the game.

Example: Key in 45. Have students guess, then push ⊜. The result should be **44**.

If the game is too simple, use starting numbers of 90, 110, or 1010. Two students can play, taking turns putting in numbers and seeing if the other can guess what will result after ⊜ is pushed.

Variations for grades 4–6. Start with 10 ⊟ ⊜ or with 100 ⊟ ⊜.

Variations for junior high. Use .1 ⊟ ⊜ or .01 ⊟ ⊜ as the starting sequence. A real stumper is .05 ⊟ ⊜.

AFTER

"After" is a variation of "Before," in which the constant is added to rather than subtracted from the number on the display. To set the calculator up, key in 1 ⊞ ⊜, or 10 ⊞ ⊜, or 100 ⊞ ⊜. Then play exactly the same as "Before."

Example: Key in 1 ⊞ ⊜. Then key in 45 and have students guess the result before the ⊜ is pushed. The result should be **46**.

SOLITAIRE

Teaches: Basic mathematics facts and calculator functions

For: One player and a calculator

Object of the game: Using only the legal keys, to get the goal number on the display.

How to play: Pick a goal number and a set of "legal" keys. The game is to see who can get the answer in the fewest keystrokes.

Example: Goal number, 17, and legal keys 5, ⊞, ⊟, ⊠, ⊡, or ⊜. Using only these keys, get **17**. One answer would be

5 ⊞ 5 ⊡ 5 ⊜ ⊞ 5 ⊞ 5 ⊞ 5 ⊜.

This solution takes 13 keystrokes. There are shorter solutions.

Pick different goal numbers or vary the legal digit.

Variations for primary grades. Use goal numbers that are multiples of the legal digit.

Example: Using 2 and the keys ⊞, ⊟, ⊠, ⊡, or ⊜, get **24**. A first grader might do 2 ⊞ 2 ⊞ 2 ⊞ 2 . . . , until he got to **24**. A shorter solution would be 22 ⊞ 2 ⊜.

Variations for junior high. Use negative goal numbers.

Example: Using 5 and ⊞, ⊟, ⊠, ⊡, or ⊜, get **−3**.

Sophisticated solutions can be implied.

Example: Using 5 and ⊞, ⊟, ⊠, ⊡, or ⊜, can you get to **26** in only ten keystrokes? (Try it before looking at the solution at the end of this article.)

TARGET K

Teaches: Decimal place value and bracketing guesses

For: A whole class and a calculator, or a small group or single student and a calculator

Object of the game: To get the target number on the display. The decimal part of the target number doesn't count. For example, if the target is **500**, a display of **500.4716** is correct.

How to play: Put a multiplication constant into the machine. (Warning: Do not clear the machine after the constant has been put in, otherwise the constant will disappear.) The object of the game is to guess what number, when multiplied by the multiplication constant, will give the target number.

Example: If the target number is 500 and the multiplication constant is 17, then key in 17 ☒ ⊟. Make a guess, say 50 ⊟, which gives 850—too large. Make another guess, say 20 ⊟, which gives 340—too small. The correct answer must be between 20 and 50. When students find that 29 ⊟ is too small and 30 ⊟ is too large, they will begin trying decimals.

Variations for primary grades. Limit the target numbers to exact multiples of the constant. For example, the target number could be 265 and the constant 5 ☒ ⊟. Still simpler is a target of 24 and a constant 8 ☒ ⊟.

Variations for junior high. Pick a small target number and a large constant.

THE BIG ONE

Teaches: Decimal place value

For: A small group or a single player and one calculator

Object of the game: To get the display to read 1

How to play: One person sets up the calculator for the player(s) without the players seeing what was put in the calculator. The calculator is set up by keying a number, then ⊞ ⊟. After the calculator is set, players guess numbers and push ⊟ until the display is 1.

Example: Select a "mystery" number between 1 and 100—say 27. Put it into the machine by keying 27 ⊞ ⊟. Pushing the ⊞ ⊟ keys makes the number 27 a constant divisor. (Caution—do not clear the machine at any time after the mystery number has been put in or the number will be wiped out.) Give the calculator to the players, who try to guess the mystery number by keying in trial numbers and then pushing ⊟. When the mystery number is guessed, the display will show 1 after the ⊟ key is pressed. If the number tried is not correct, the display gives clues by showing what the guess divided by the mystery number equals.

Variation for primary grades. Limit the mystery numbers to between 1 and 10.

Variation for junior high. Use three-digit mystery numbers.

Those are the games. Here are a few tips on how to introduce them to a class.

Probably the best introduction is to play the game first with the entire class, or to play it with one person in front of the entire class. Even though the class cannot see the tiny digits on the calculator display, you can read the numbers out loud to them. This gives the students a feel for the game that they cannot get by just reading the directions at an interest center. After the rules have been explained and they have seen the game played once or twice, then let the students play in pairs or in small groups. After they have done that for five or ten minutes, students should be able to play the game independently. If you show them the game without letting them play it immediately, they have usually forgotten the rules or lost interest by the time their turn at the interest center comes.

Although the kids get pleasure from playing any of these games just once or twice, the real benefits accrue to a student after playing the same game a number of times. Strategies become more sophisticated and generalizations develop. So introduce a single game and let the students play it briefly but regularly over a two- to three-week period. Generate enthusiasm through contests, posted problems about the game, or championship. Then introduce another game. This strategy allows these games to be real teaching aids, rather than simply amusing pastimes.

NOTE. The solution to the "Solitaire" puzzle is

5 ☒ ⊟ ☒ ⊟ ⊞ 5 ⊞ 5 ⊟.

Ideas

Prepared by **Earl Ockenga**
and **Joan Duea**
Malcolm Price Laboratory School
University of Northern Iowa
Cedar Falls, Iowa

Each IDEAS presents activities that are appropriate for use with students at various levels in the elementary school. The activity sheets are so arranged that they can be easily removed and reproduced for classroom use. Permission to reproduce them for such use is not necessary.

The IDEA this month is to use a hand-held calculator to provide practice in estimation.

IDEAS — For Teachers

Objective: Experience in addition

Levels: 1, 2

IDEAS — For Teachers

Objective: Experience in estimating sums and differences

Levels: 3, 4, 5

IDEAS — For Teachers

Objective: Experience in estimating products

Levels: 5, 6, 7

IDEAS — For Teachers

Objective: Experience in estimating quotients

Levels: 7, 8

Directions for teachers:

The way you use the activities will depend on the number of calculators available to you.

If you have one calculator:

1. Remove the master copy and reproduce a transparency for use with an overhead projector.
2. Separate the students into two teams (team *X* and team *O*).
3. Project the transparency. Tell students there will be a five-minute warm-up before playing the game. During the warm-up session, students are to use their estimating skills to identify pairs of numbers whose answers (sums, differences, products, or quotients, depending on the game) are found on the game board.
4. To play the game, have the teams take turns selecting two numbers and using the calculator to compute the answer.
5. Each team finds its answer on the game board and puts the team's mark on it (*X* or *O*). The game is won when a team has an unbroken path of marked answers that connects its two sides of the game board (fig. 1).

If you have more than one calculator:

1. Remove the master copy and reproduce the worksheet.
2. Separate the students into teams, with two teams sharing a calculator.
3. Have teams take turns selecting two numbers and using the calculator to compute the answer.
4. Each team finds its answer on the game board and puts the team's mark on it (*X* or *O*). The game is won when a team has an unbroken path of marked answers that connects its two sides on the game board.

Comments:

Play the game more than once. At first students may pick pairs of numbers at random, but as they play more often they will start to develop strategies for using their estimation skills to select the numbers.

An interesting modification of the game is to require one player to pick the first number and another player on the same team to pick the second number.

Fig. 1

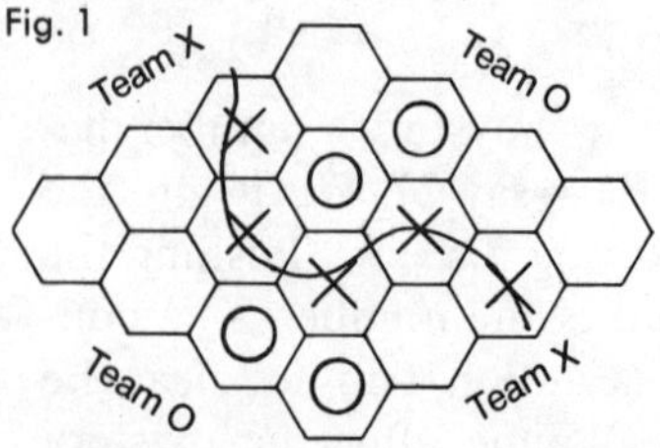

IDEAS

Addition Game (2 teams)

How to play:

1. Teams take turns. Pick any two of these numbers

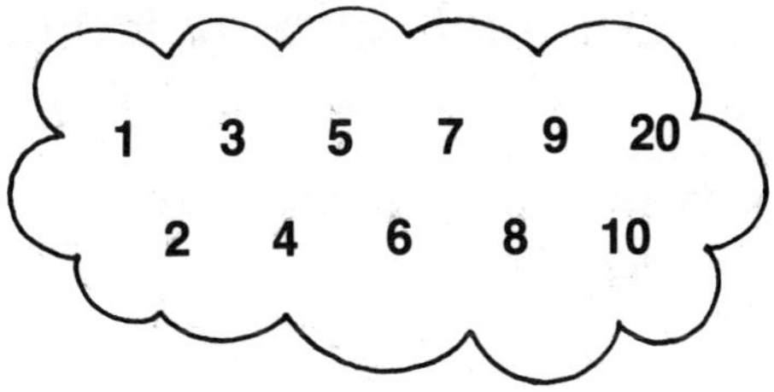

2. Add the numbers you picked.

3. If the answer is on the game board, mark it with an X or O.

Game Board

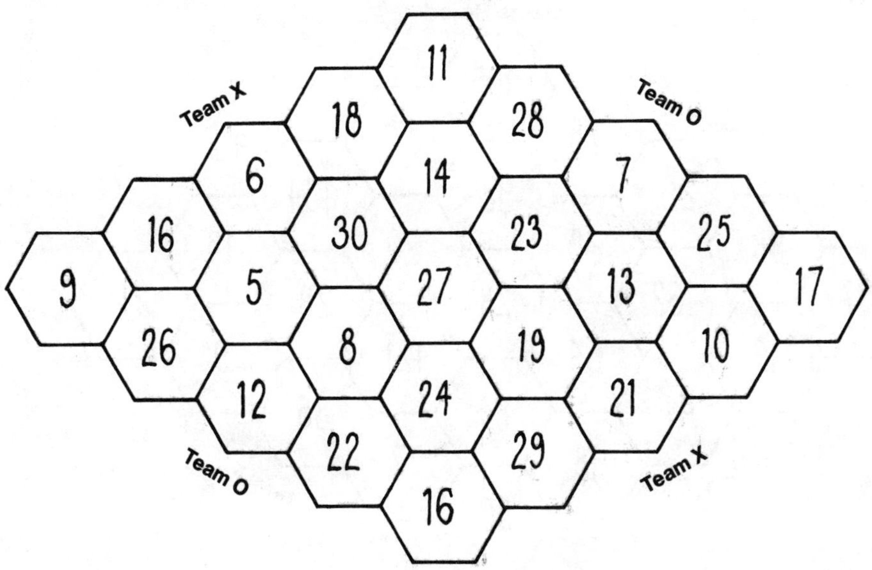

How to win: The first team to get a path of answers connecting its two sides of the game board wins.

IDEAS

Add or Subtract Game (2 teams)

How to play:

1. Teams take turns. Pick any two of these numbers.

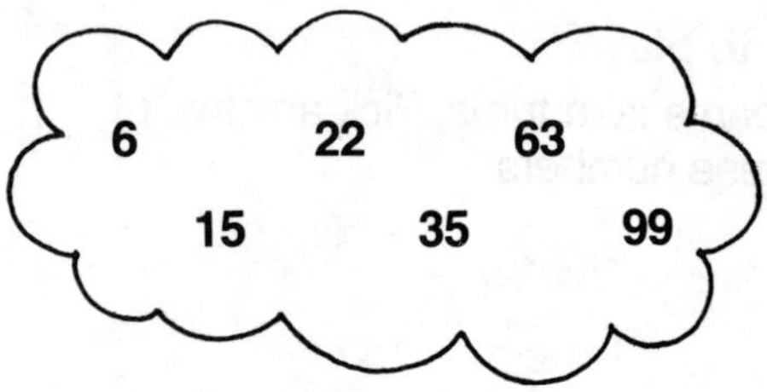

2. Add or subtract the numbers you picked.

3. If the answer is on the game board, place your team's mark on it (X or O).

Game Board

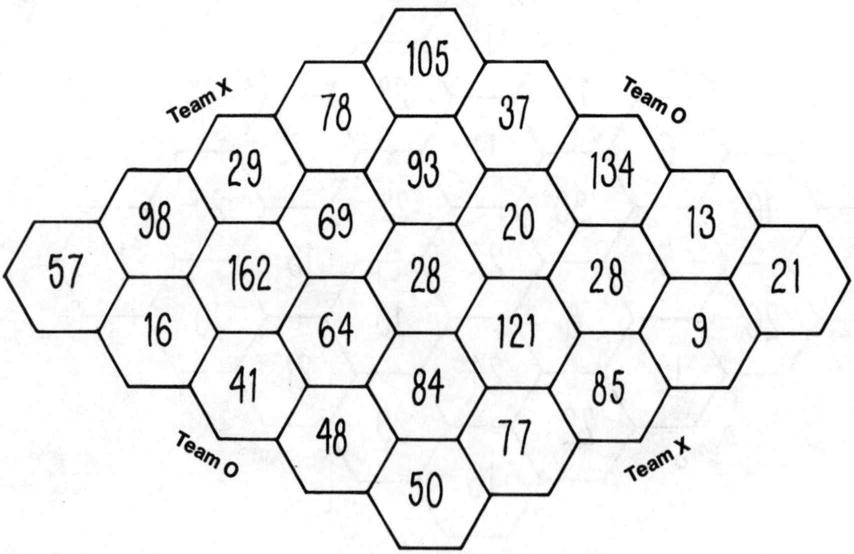

How to win: The first team to get a path across the game board wins.

IDEAS

Multiplication Game (2 teams)

How to play:

1. Teams take turns. Pick any two of these numbers.

11 31 51 71 91

21 41 61 81

2. Multiply the numbers you picked.

3. Find the answer on the game board. Place your team's mark on it (X or O).

Game Board

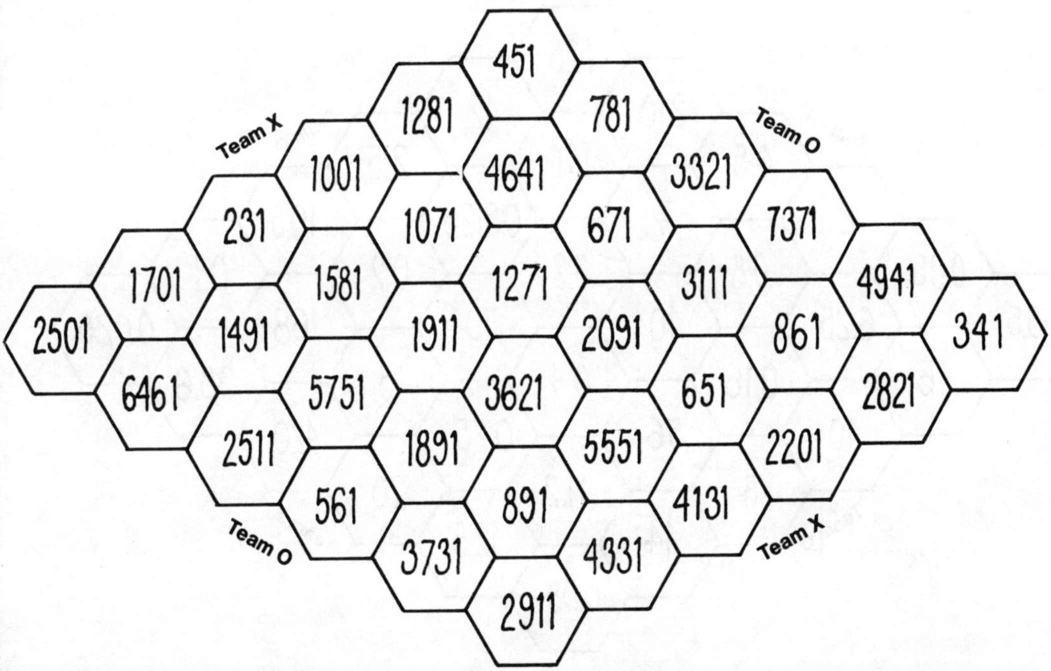

How to win: The first team to get a path of answers connecting its two sides of the game board wins.

IDEAS

Division Game (2 teams)

How to play:

1. Teams take turns. Pick any two of these numbers.

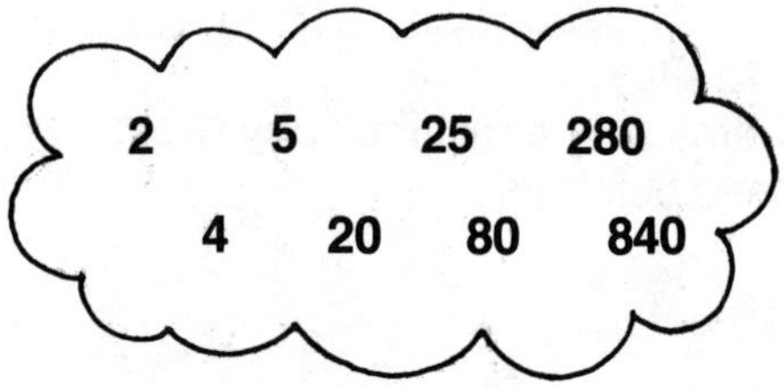

2. Divide the numbers you picked

3. If the answer is on the game board, place your team's mark on it (X or O).

Game Board

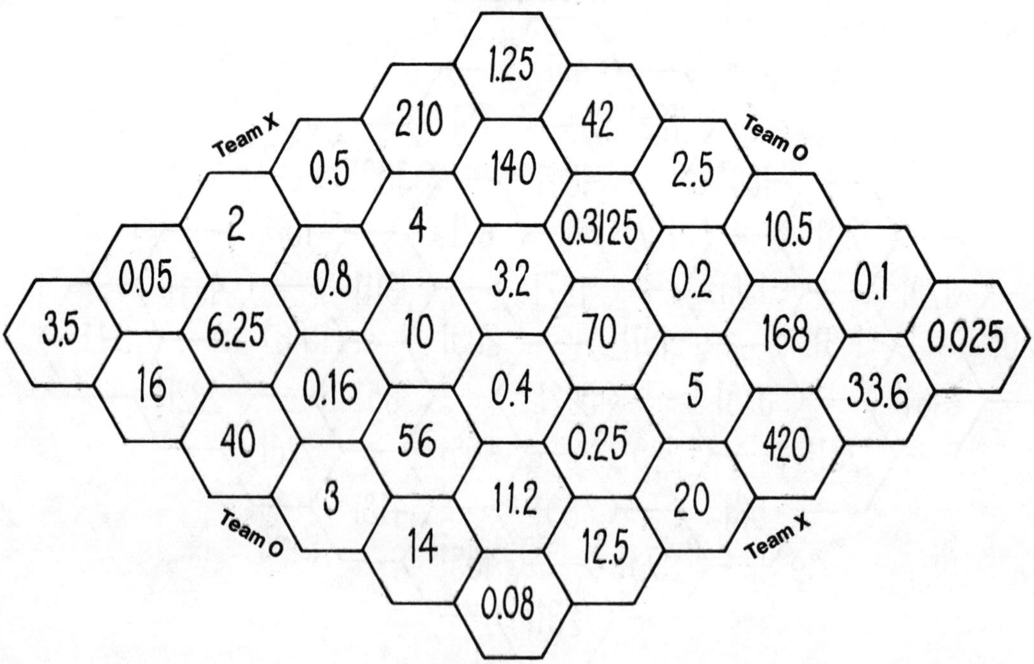

How to win: The first team to get a path of answers connecting its two sides of the game board wins.

Check your calculator computations

CHARLES D. FRIESEN

Currently chairman of the mathematics department at Lincoln High School in Lincoln, Nebraska, Charles Friesen has also had teaching experience at the junior high and college levels. For two years he also supervised mathematics student teachers at the University of Nebraska—Lincoln.

Perform the indicated computations on a calculator. To check your answer turn the calculator upside down and read a word answer. A clue is given for each problem.

Calculation	*Numerical answer*	*Clue*	*Word answer*
1. $(160 \times 5) + 7$	____	A tennis shot	____
2. $456 \times 81 - 1828$	____	The capital of Idaho	____
3. $33\,624 \div 6$	____	Mature pigs	____
4. $394\,752 \div 512$	____	Sick	____
5. $2715 + 330$	____	Worn on the foot	____
6. $584\,831 - 123\,456$	____	Vehicle for riding on snow	____
7. $8^3 + 2$	____	Not hers, but ____	____
8. $12\,345 + 23\,456 - 465$	____	Their message is "Honk, honk"	____
9. $7 + (50 \times 110)$	____	Not a gain, but a ____	____
10. $((654\,321 \div 3) - 214\,238) \times 2$	____	It rings	____
11. $5787 \div 3 \times 4$	____	Fish organ	____
12. 22% of 2900	____	To plead	____
13. $17^2 + 7^2$	____	It buzzes about	____
14. $706 - 99$	____	A ____ cabin	____
15. $1884 + 1623$	____	To fail to win	____
16. $64\,118 - (80)^2$	____	Letters that arrive near the first of the month	____
17. $279^2 - (16 \times 31)$	____	The outer covering of a peanut	____
18. 15×247	____	The bottom of a shoe	____
19. $987\,654 - 984\,150$	____	Firetruck equipment	____
20. $59^2 + 223$	____	A pit	____

Let's Do It!

By **James V. Bruni** *and* **Helene J. Silverman,**
Herbert H. Lehman College, City University of New York

Taking advantage of the

If you have experienced the excitement of a group of children who realize they will be using a hand calculator, you will agree that the calculator *should* have a role to play in the elementary school classroom. Instant motivation! The most "reluctant" learner is anxious to have a chance to use the calculator.

The important question is, *How* should children be using the hand calculator? Parents and teachers alike are often hesitant to have children use a calculator for fear they will become dependent on the use of the calculator and *never* develop essential computational ability. That is certainly an important concern, but the judicious use of a calculator presents many rich possibilities for enhancing elementary school mathematics programs. Since the relatively easy availability of calculators is such a recent phenomenon, teachers have only begun to explore these possibilities.

The activities that follow suggest some ways for using the hand calculator as a complement to any mathematics program. They focus on developing skills for using the calculator, exploring basic arithmetic operations, understanding algorithms, mental calculation and estimation, and problem solving. Most activities can be done individually and in small groups and require the availability of only one or two hand calculators.

GETTING TO KNOW THE CALCULATOR

As with the introduction of any new material, it is important that the teacher spend some time getting acquainted with the calculator. Can you make it work? What do the different buttons do? Give the children a chance to *play* with the calculator, too, before introducing more structured activities. Encourage them to discuss their discoveries with each other. You might have them make up problems and try them out on the calculator, or use it to check their arithmetic homework.

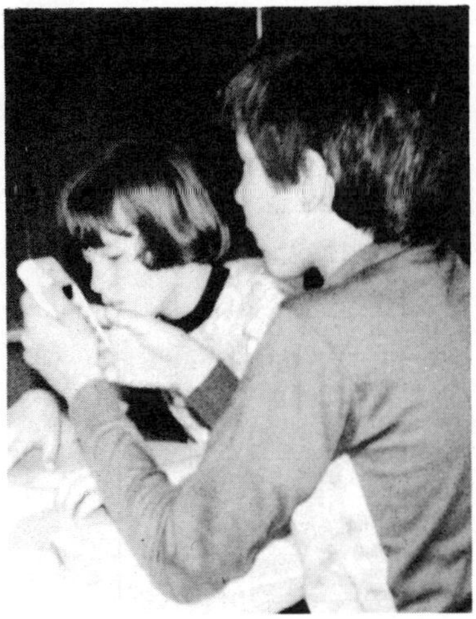

Photographs by Clif Freedman

hand calculator

Very often through free play with the calculator (or after a carefully placed leading question on your part), children discover that some interesting things can happen. For example, when they work out a problem like 81 ÷ 3 on the calculator and continue to depress the [=] button repeatedly, each successive quotient is divided by 3.

If you look at the ways children choose to use the calculator and listen to their comments, you may pick up valuable clues for worthwhile activities. For example, a child announces that he has figured out his age on the calculator. You might ask him how he did it, or other children might suggest how it was done. You may wish to take advantage of this apparent interest and structure a problem like the following:

If you are 9 years old in 1976, how old will you be in

1980? ____
1983? ____
1990? ____
2000? ____
2025? ____

Choose a month for each of these years. Can you figure out how many *months* old you will be?

BASIC OPERATIONS WITH THE CALCULATOR

Once children have had the opportunity to learn to operate the calculator, you may wish to introduce task cards like the following to help them become familiar with performing successive operations on the calculator. At first the final answers can be included on the card. Then, the children can find the final answers and compare results with a partner, or the task card can be self-checking.

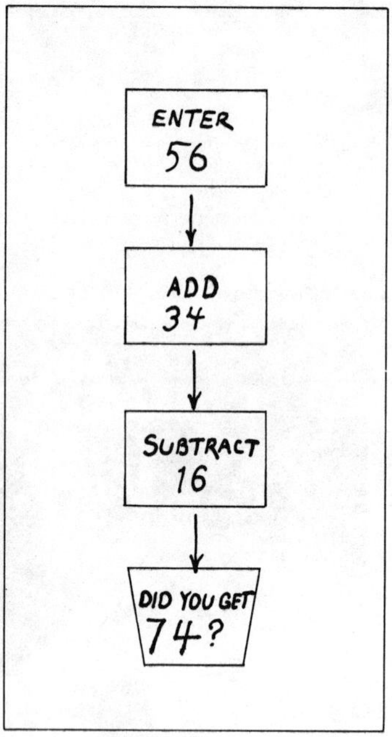

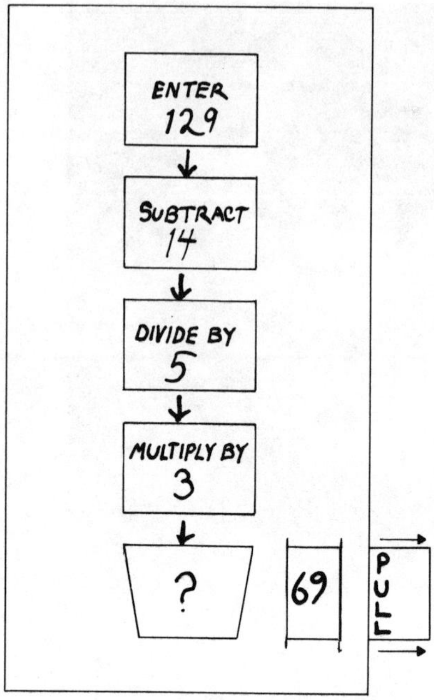

You can develop some interesting problem-solving situations by omitting a number other than the final answer. How do the children solve the problems? What relationships do they see between operations?

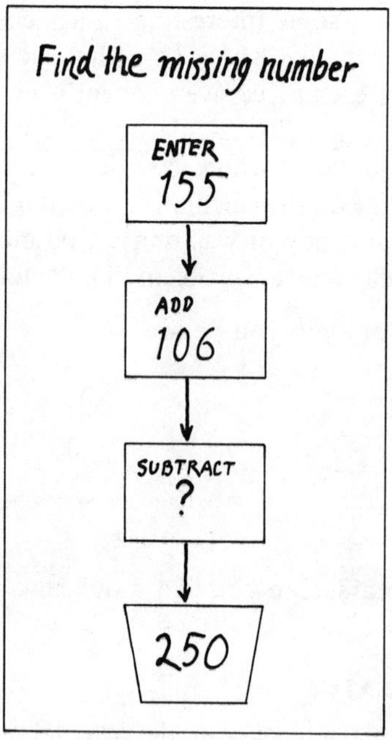

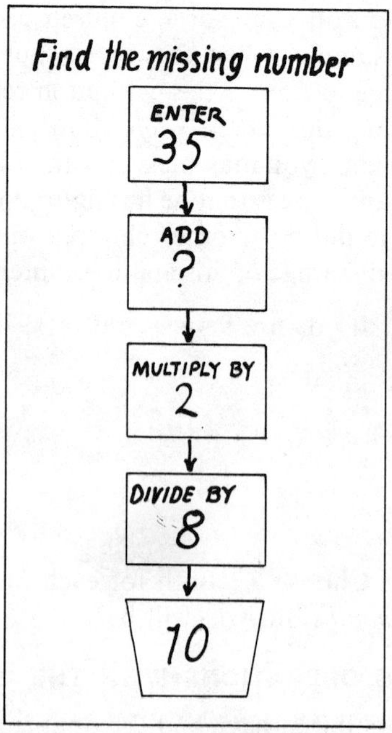

In a similar way, you can make more open-ended task cards, where many solutions are possible. How many different ways can the children make the following work?

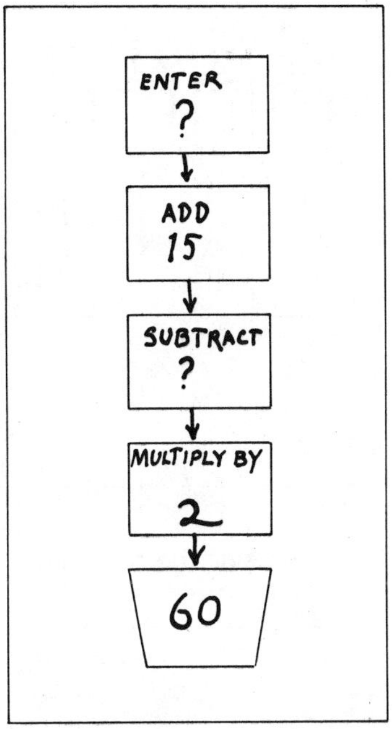

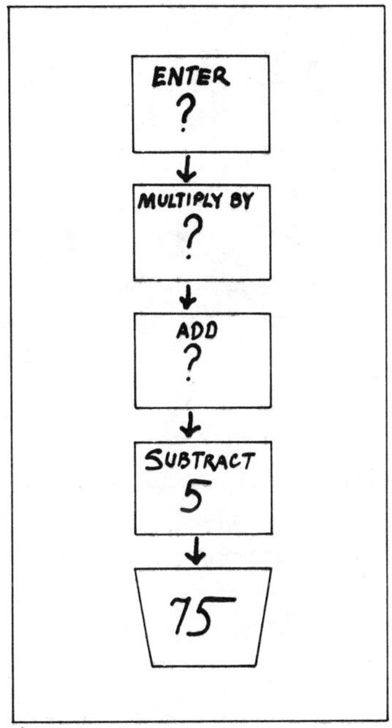

If this activity is enjoyable, you might suggest that children make similar task cards of their own for others to solve, using the calculator to help set up the problem and to verify solutions.

ANALYSIS OF ALGORITHMS

Rather than serving as a threat that children will never learn basic algorithms, the calculator can prove to be a useful tool for helping children understand those algorithms. Activities like the following can lead children to take a closer look at why an algorithm works or to suggest alternate ways for performing an arithmetical operation:

Find the missing numbers. Why do they work?

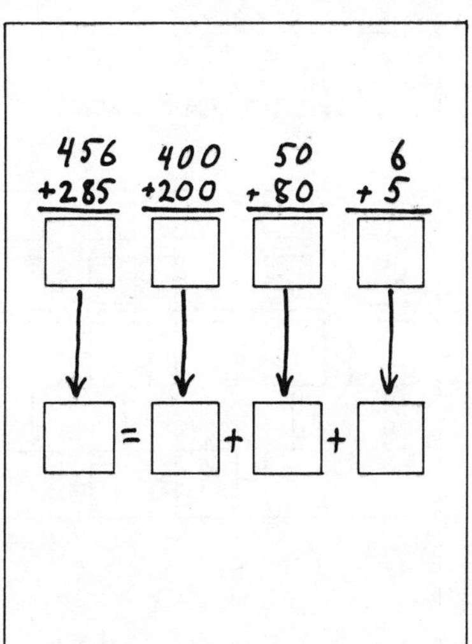

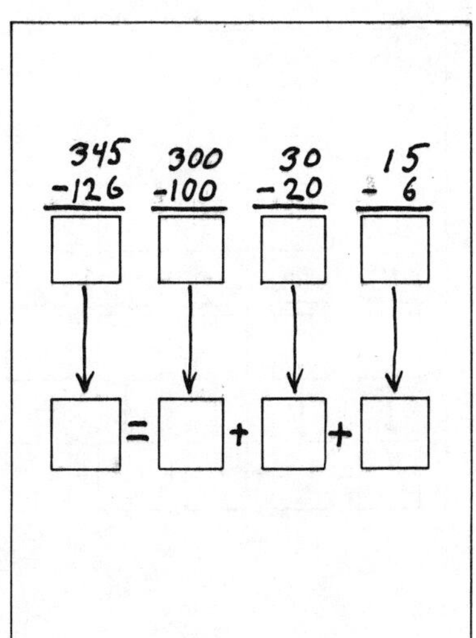

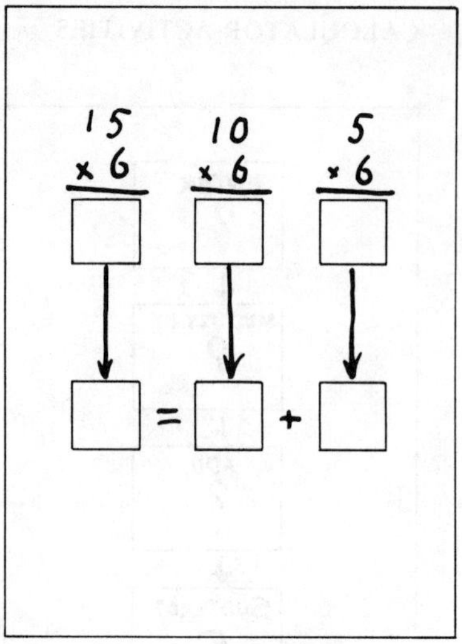

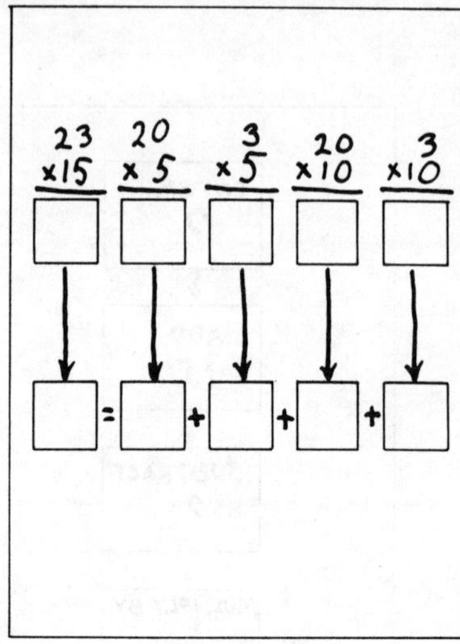

Can you make these work? Check by using the calculator.

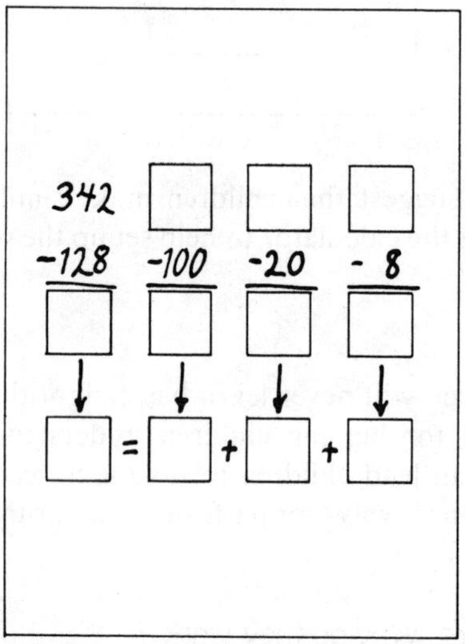

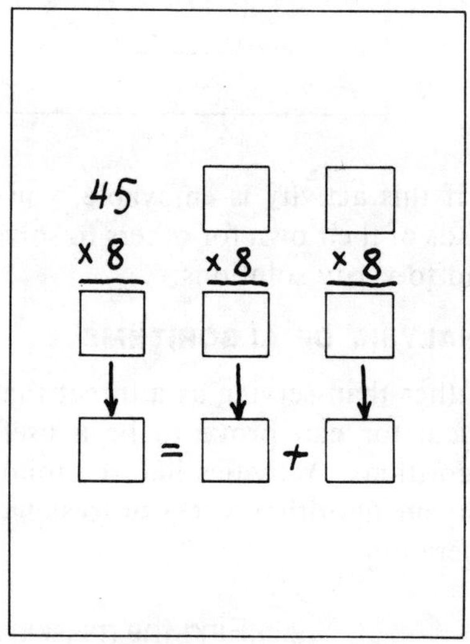

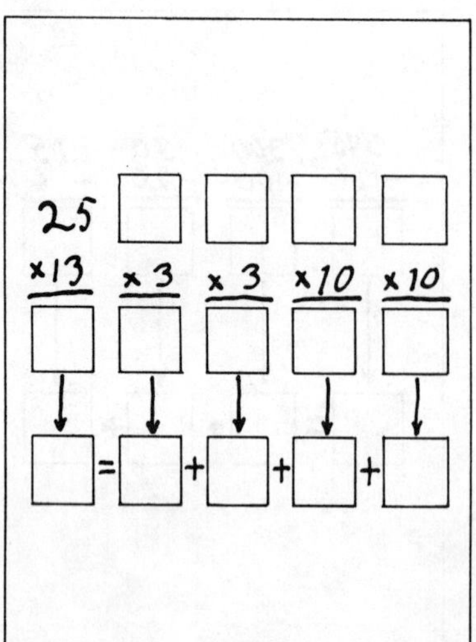

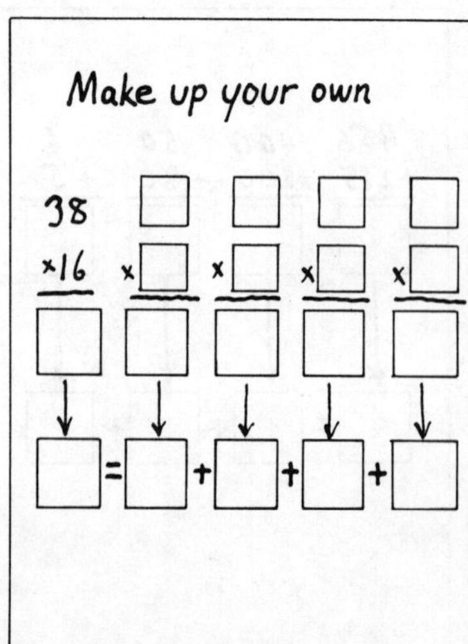

ESTIMATION AND MENTAL CALCULATION

One area of the mathematics program that is often neglected is estimation and mental calculation. You can invent games like the following that make use of the calculator and help children develop those important skills.

How close can you get?

Participants. Four players and one or two persons serving as "judges." Each judge has a hand calculator.

Materials needed. A set of fifty or more cards numbered 1, 2, 3, . . .

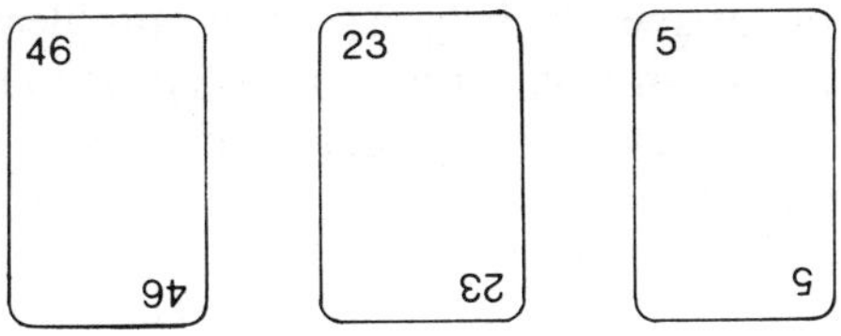

A set of ten "key" cards numbered 50, 55, 60, 65, 70, 75, 80, 85, 90, 95, and 100.

Procedure. The game is played as follows:

1. Each player is dealt five cards from the set of fifty cards.
2. The stack of key cards is placed in the center of the playing table and one key card is turned over. The number on that key card will be the key number for that game.
3. Each player selects *three* of the five cards he was dealt, attempting to select the three numbers that when added have a sum closest to the key number. Players make their selections by estimating or by mental addition.
4. After the selections are made, each player puts his three cards on the playing board in front of him for all to see.
5. Each player then has the opportunity to look at the cards that the other players have selected. If a player feels that his sum is closer to the key number than the sum of any other player, he stays in the game. Otherwise, a player turns his cards over and is out of the game.
6. Meanwhile the judges use the calculators to calculate the sum for each player and announce the winner.

Scoring. The winner gets one point for having the closest sum, plus one point for each player who stayed in the game. In case of a tie, the points are shared. The player with the most points at the end of the session is the winner.

Variations. Many variations are possible. Players might be allowed to keep from one to five of the cards dealt to them. Or, the cards might include larger numbers, negative numbers, decimals, and so on, according to the ability of the players.

DOING WORD PROBLEMS

Sometimes when children first use a calculator they are so impressed with its "magic" they feel they can do *any* mathematics problem if they can only be allowed to use the calculator. Unfortunately, their conception of a mathematics problem is often limited to solving some number problems like the following:

$$\begin{array}{r} 145 \\ +\,324 \\ \hline \end{array} \quad \text{or} \quad \begin{array}{r} 328 \\ \times\,17 \\ \hline \end{array}$$

It can be very worthwhile to provide situations where children get the opportunity to use the calculator to help solve *word* problems. This can help them realize that, while the calculator is a tremendous tool, it does not help in solving a problem unless you know what numbers you want to use, in what order, and what operations to perform. The following game can be played using word problems of varying degrees of difficulty (as appropriate for the players).

Can you solve it?

Participants. Six persons forming two teams of three players. Each team has one person who must do the problem mentally, a second person who can use paper and pencil, and a third person who can use a hand calculator. One person (not a member of the teams) serves as a leader.

Materials needed. Players will need—

(*a*) small slips of paper and pencil for each player
(*b*) one hand calculator for each team
(*c*) chalkboard or record sheet
a collection of word problems based on real-life situations and appropriate for the interest and ability of the players. Menus, catalogs, advertisements, and so on can provide information for problems. Each problem is on a separate card.

Examples:

Suppose you buy two shirts for $5.99 each.
How much change would you get from $20.00?

How many bikes that cost $55.00 can you buy with $500.00?
How much money would you have left over?

Procedure. The game is played as follows:

1. A problem is read by the leader. Each player attempts to solve the problem, without consulting with teammates, and records his answer on a slip of paper.
2. The leader collects the answers and records them on the chalkboard.

Example:

Answers

Team A		*Team B*	
M. C.	45	M. C.	43
P. P.	36	P. P.	45
H. C.	48	H. C.	45
Team	_____	Team	_____

M. C. = Mental calculation P. P. = Paper and pencil H. C. = Hand calculator

If answers vary, each team gets an opportunity to discuss the answers and comes to a consensus. Team answers are then recorded.

Scoring. Points are awarded for all correct answers as follows:

Mental calculation	3 points
Paper and pencil	2 points
Hand calculator	1 point
Team answer	1 point

The team that accumulates the most points wins.

The preceding activities only begin to suggest ways you can make use of the hand calculator in the elementary classroom. Used judiciously, the calculator can be an invaluable asset to any mathematics program.

Problem Solving with Calculators

By **Janet Parker Morris**

Any teacher who has experienced the excitement and complete concentration of a group of children working with calculators knows that the question is no longer *whether* calculators should be used in elementary schools, but *how* they should be used. Ways to use calculators with existing texts and mathematics programs are being sought by teachers at all grade levels. Calculator activities that require problem-solving skills, that use the calculator as a tool to avoid "getting bogged down" in tedious computations, and that stimulate the imagination and interest of students, are valuable and compatible with existing programs.

Several such activities that use basic four-function calculators are described in this article. Each activity is classified under a heading that suggests a specific way in which the calculator is used to develop skill in problem solving. Many of the activities develop abilities other than those under which they are listed since all the activities use the calculator in basically the same way—as a calculating device that frees the user from the burden of the computations that are involved in order that he or she can concentrate on the higher-level aspects of the problem situation. In each activity, the calculator is used only as a machine that does the computing; the student user does the thinking.

Use the calculator to explore number patterns

Freed from the burden of computational drudgery, students find number patterns intrinsically interesting, and studying number patterns adds insight into working with numbers.

An assistant professor of mathematics education at the University of Michigan in Flint, Janet Morris is currently interested in a research project that involves working with children in area classrooms to implement the use of calculators in classroom instruction. She has also conducted an experimental course for inservice teachers on calculators as an instructional aid.

Activity. Use your calculator to find the first few of the following products, then predict what the others are. What patterns did you use for your prediction?

$9 \times 12345679 =$
$18 \times 12345679 =$
$27 \times 12345679 =$
$36 \times 12345679 =$
$45 \times 12345679 =$
$54 \times 12345679 =$
$63 \times 12345679 =$
$72 \times 12345679 =$
$81 \times 12345679 =$

(Teaching note: Use the automatic constant capability of the calculator to eliminate tedious keying in activities such as this.)

Activity. What do you notice in the following exercises? Will the pattern continue on and on, or will it end? If it ends, when will it end?

$11 \times 11 =$
$111 \times 111 =$
$1111 \times 1111 =$
$11111 \times 11111 =$
.
.
.

$9 \times 9 =$
$99 \times 99 =$
$999 \times 999 =$
$9999 \times 9999 =$
.
.
.

Activity. Can you find the pattern after using your calculator on only the first three of the following examples? Predict the results for the remaining examples, then check your predictions with the calculator.

$101 \times 222 =$
$101 \times 2222 =$
$101 \times 333 =$
$101 \times 3333 =$
$101 \times 33333 =$
$101 \times 555 =$

Activity. Multiply several two-digit numbers by 99. Study the products, then write down three things that you notice about the products. Multiply several two-digit numbers by 999. Again, study the products, then write down three things that you notice about the products. (Teaching note: Until students are familiar with such open-ended problems, you might want to ask more directive questions. For example, How do the digits of the product compare with those of the two-digit number? or When you add the digits of the product, what do you notice? Also, whenever feasible, ask the more experienced students to explain why these patterns occur. Merely suggesting that the students think of 99 as $100 - 1$ usually spurs informal but remarkably keen explanations.)

Activity. In figure 1, use your calculator to find the decimal names for the fractions in the first part. Then use the patterns that you notice in these to predict the decimal names for the fractions in the second part. Check your predictions with the calculator.

Fig. 1

Find these using the calculator:

$\frac{1}{11} =$	$\frac{2}{11} =$	$\frac{3}{11} =$
$\frac{1}{111} =$	$\frac{2}{111} =$	$\frac{3}{111} =$
$\frac{23}{11} =$	$\frac{36}{11} =$	$\frac{48}{11} =$

Use what you just found to predict these:

$\frac{8}{11} =$	$\frac{9}{11} =$	$\frac{10}{11} =$
$\frac{7}{111} =$	$\frac{9}{111} =$	$\frac{5}{111} =$
$\frac{59}{11} =$	$\frac{78}{111} =$	$\frac{63}{11} =$

Activity. The product in the following multiplication is too big for the calculator. Can you find the product by using the calculator with smaller numbers and finding a pattern that will lead to the answer of this problem?

101 × 77777777 =

Use the calculator to discover relationships and to develop concepts that may be obscured by tedious or difficult computations

Activity. With a centimeter tape, find the circumference and the diameter of five different circular objects that you see in the room. Record these measures in a chart like the following one. Then use your calculator to find $c \div d$ and complete the chart. What do you notice about the entries in the last column? What would you predict is true of any circle?

Object	Measure of circumference c	Measure of diameter d	$c \div d$

Activity. Use a centimeter ruler to draw three different right triangles on centimeter graph paper. Measure each side and record the length in the following chart (c is the longest side). Then use your calculator to complete the chart. Make a statement about what you discover.

	Measures of sides						
Triangle	a	b	c	a^2	b^2	c^2	$a^2 + b^2$
1							
2							
3							

Use the calculator to practice mental estimation

The essential skills of mental estimation and checking answers for their reasonableness are difficult to teach and are often neglected. Skill in making good estimates is important with calculator work in order to catch keying errors.

Activity. Use estimation to locate the place for the decimal point in each answer in the following examples. Then use your calculator to check to see if you were correct.

6.7 + 5.2 + 9.6 = 2 1 5
47.58 − 8.71 = 3 8 8 7
23 × 4.2 = 9 6 6
74.23 × .061 = 4 5 2 8 0 3
56.925 ÷ 0.9 = 6 3 2 5

Activity. Use number properties, rounding, and estimation to determine which sign, >, <, or =, goes in each □ to make each of the following statements true. Then use the calculator to check your accuracy.

19 × 31 □ 19 × 32
3 × 40 × 80 □ 3 × 39 × 60
47 × 99 □ 47 × 100 − 47
669 − 296 □ 679 − 296
248 ÷ 15 □ 248 ÷ 12
625 ÷ 5 □ 625 ÷ .5

Activity. Use estimation and your calculator to help you decide which of the prime numbers, 2, 3, 5, 7, 11, 13, 17, or 19 are the factors that make the following multiplication statements true. Would problems like these be harder if composite numbers were used? Why?

___ × ___ = 91
___ × ___ × ___ × ___ = 910
___ × ___ × ___ = 935
___ × ___ × ___ = 1001

Activity. Use your calculator to help you determine which sign of operation (+, −, ×, or ÷) goes in each □ to make the following statements true. How does estimation help you decide which operation sign to try?

(73 □ 26) □ 23 = 2277
(62 □ 21) □ 236 = 1066
1776 = (882 □ 49) □ 1758
215 □ 896 □ 788 □ 412 = 735

Use the calculator to reinforce the inverse relationships of operations and provide readiness work for equation solving

The concept that to solve an equation means to find a value that may replace the unknown is enhanced since the calculator makes the "check by substitution" method immediately available.

Activity. Find the missing factor in each of the following examples by solving the related division problem with your calculator. Check your answer using multiplication.

17 × ___ = 1632
236 × ___ = 14868
0.78 × ___ = 12.168
6.3 × ___ = 3754.8

Activity. Use your calculator to complete the following multiplication table.

×		23	
98	4606		
64			5440
		299	

(Teaching note: Such tables can be made more challenging by having one entry incorrect. Students must then determine which entry is wrong in order to uniquely complete the table.)

Activity. Puzzle: I am thinking of a number. I divide it by 16, then multiply the answer by 47, then subtract 19 from that answer, then add 7, and I end up with 1069. What was my original number?

Make up a problem like this for a friend.

Activity. If $a \times b = 1495$ and $b = 65$, what must a be? What must $a + b$ be? Enter this information in the first line of the following chart. Find other values of a and b to complete the other lines of the chart.

a	b	$a \times b$	$a + b$
	65	1495	
47			66
		91	20

Use the calculator in application problems

Because the calculator handles "messy numbers" and large numbers easily, problems that are more realistic and interesting than standard "textbook" story problems can be used. Such problems can even be used with students who have not yet completely mastered the computational algorithms, since the calculator allows the focus to be on *which* operation to use in a given situation rather than *how* to do the computation.

Activity. Make a "personal information card." Find out data about yourself to include on your card—How many minutes have you lived? How many hamburgers would you eat in ten years at your estimated weekly rate? How many times does your heart beat in a year? And so forth.

Activity. Find three boxes in the classroom. Estimate their volumes. Then measure the boxes and compute their volumes. Was your error on each box within 15 percent of the box's volume?

Activity. If your rich uncle gave you 1¢ the first day, 2¢ the second day, 4¢ the third day, and so on, doubling each day, how many days would it take for you to be a millionaire?

Use the calculator to develop the problem-solving technique of "guess-then-check"

Famous and successful problem solvers such as Polya advocate this technique, yet it is frequently neglected because the computations that are involved in the check are often prohibitive. Also, too many students have the attitude that it is wrong to ever make a guess. Actually, one can learn from a guess by checking it.

Activity. In the following examples, guess the number that goes in each □. Check your guess with the calculator, then guess again, improving your guess. Which numbers can you find within five guesses?

$$\square^3 = 12167$$
$$\square^3 = 704969$$
$$\square^4 = 1419857$$
$$\square^5 = 16807$$

Challenge: Find the solution to $\square^3 = 3$ good to five decimal places.

Activity. The division of $1 \div 8$ gives a quotient of 0.125. Can you do a division that gives a quotient of 0.775?

Use the calculator for individual exploration and enrichment

Student enthusiasm for calculator games can provide much valuable practice and learning if games are chosen that require some type of problem solving in order to make wise moves and find winning strategies. After students play a game several times and develop more sophisticated strategies, the game becomes a teaching aid rather than simply an amusing pastime.

Activity. Use the 8 key eight times and the + key an unlimited number of times to get a display of 1000.

Make up a problem like this for a friend.

Game. Play "21" with a friend. Alternate turns entering the number 1, 2, or 3 and pressing the + key. The first person to get a display of 21 wins.

Can you find a way to always be the winner? Does your winning strategy work if different digits are allowed? If the target number is changed?

Game. Play "Reverse." Have a friend enter any three-digit number into the calculator. Your goal is to get as close as possible to the number formed by reversing the digits of this starting number. If the start number is smaller than the goal number, the operation you may use is addition; if it is larger, the operation is subtraction. The numbers you may use must be two-digit numbers formed from the digits currently on display. For example, if the starting number is 246, then your goal is 642 and the operation is addition. Your first play is to add any two-digit number composed of 2, 4, or 6. Suppose you add 62. The display then becomes 308, and for your next move you may add any two-digit number formed from 3, 0, or 8. Continue play until your sum is as close to the goal number as possible. Record your closing sum. Your friend then begins with your goal number and tries to reach the original number, following the same rules. The player getting closest to his goal number wins.

After you have played this game several times, make up a game like it with different rules. You may want to try larger numbers or different operations.

Game. Play "Erase." Have your friend enter any six-digit number into the calculator. Now see what is the fewest number of moves it takes you to get a display of zero. For each move, you may add, subtract, multiply, or divide by any two-digit, nonzero number. Then trade places and have your friend try to "Erase" a six-digit number you enter into the calculator.

What did you notice about numbers that makes them easy or difficult to erase?

These activities illustrate ways in which calculators can be used now, with existing programs and curriculums. Mathematics educators generally agree that calculators will dramatically affect the content and methodology of school mathematics. Long-term, substantial curriculum changes, however, should be slow in coming, the result of thoughtful evaluation of research studies. In the meantime, teachers can capitalize on the potential of calculator work with activities such as these.

These activities are just a few of the ways calculators can be incorporated into existing curriculums to develop problem-solving skills. As calculators are used more widely, more ways will become apparent to teachers and to students. Some excellent activities have originated with students investigating and experimenting with calculators. Teachers should never underestimate the range of problems their students will undertake with calculators when the students are encouraged to "try to find out for yourself."

Acknowledgments

Ideas for these calculator activities have been gathered from a wide variety of sources, as well as shared and modified with other teachers and students, making it impossible to give specific references. However, having heard them speak on a number of occasions and read many of their works on calculators, this writer is indebted to George Immerzeel (University of Northern Iowa) and Wallace Judd (Stanford University) for many excellent ideas. Teachers will find their publications invaluable sources of calculator activities.

References

Immerzeel, George. *'77 Ideas for Using the Rockwell 18R in the Classroom.* Foxborough, Mass.: School Images, 1976.

Judd, Wallace. *Games, Tricks, and Puzzles for a Hand Calculator.* Menlo Park, Calif.: Dymax, 1974.

Michigan Council of Teachers of Mathematics (MCTM). *Uses of the Calculator in School Mathematics, K-12.* Lansing, Mich.: MCTM, 1977.□

Minicalculators and Repeating Decimals

By **Billy F. Hobbs and Charles H. Burris**

The minicalculator can be used effectively in the classroom to encourage student inquiry and creativity through experimentation with mathematical ideas. The benefit of the calculator is multiplied when the experimentation requires computations that are too difficult or time-consuming to be done by paper-and-pencil algorithms. (See "Minicalculators in Schools," *Mathematics Teacher* or *Arithmetic Teacher* January 1976.)

The decimal representation of rational numbers offers a rich source for student investigation, yielding patterns that are both surprising and predictable (Jacobs 1975, Prielipp 1976). Of course, a calculator is essential in this experimentation for converting rational numbers to (infinite) repeating decimals. However, even the most powerful minicalculator will not produce most of the required representations directly.

For example, if the *period,* or block of repeating digits, is indicated by overlining, then

$$\frac{1}{3} = 0.\overline{3},$$

$$\frac{1}{12} = 0.08\overline{3},$$

and

$$\frac{1}{8} = 0.125\overline{0} = 0.125.$$

Also,

$$\frac{7}{17} = 0.\overline{411\ 764\ 705\ 882\ 352\ 9};$$

however, most calculators used in schools will display only six to ten digits. The computation $7 \div 17$ on an eight-digit calculator may result in the display

$$.411\ 764\ 71.$$

(Other possible displays are .411 764 70 and 0.411 764 7. Such variations in calculator displays will occur; they are based on the calculator's method of round-off and number representation. These variations do not affect the validity of the algorithm described in this article.)

The purpose of this paper is to provide an algorithm for generating on any calculator as many digits as desired in the decimal representation of the rational number N/D. The algorithm may be used whenever the number of digits displayed on the calculator exceeds the number of digits in the denominator, D. Only the operations $\div$, $\times$, $-$ are needed in the algorithm outlined. (Depending on the calculators available, readers may be able to reduce the number of entries by using certain other calculator capabilities, e.g., a memory.)

The Algorithm

Problem

To generate digits in the decimal representation of the rational number N/D.

Example. Find the decimal representation of 7/17, using a calculator with an 8-digit display; then $N = 7$, $D = 17$.

Notation: For any number A, let $\#(A)$ represent the number of digits in the number A; e.g., $\#(17) = 2$ and $\#(0.636\ 364) = 6$.

Let $B = \#(\text{calculator display}) - \#(D)$. To find #(calculator display), do the computation $7 \div 11$, and count the number of digits to the right of the decimal point. Thus, in this example:

$$B = 8 - 2$$
$$B = 6$$

Record

The digits produced by the algorithm are recorded in blocks of B digits, after an initial recording (digits to the left of the decimal point) that depends on whether or not $N < D$.

(1) If $N < D$, the initial record is 0.

(2) If $N > D$, convert N/D to a mixed number $A + (N'/D)$; then the initial record is A. The algorithm is then applied to N'/D. For example, the initial record for 71/13 is 5 and the algorithm is applied to 6/13, since

$$\frac{71}{13} = 5 + \frac{6}{13}.$$

Steps to follow

1. Write down the initial record for N/D.

   0.

2. Clear the calculator.
3. Compute $N \div D$.

   $7 \div 17 = \emptyset.411\ 764\ 71$

4. Adjoin to the existing record (on the right) the first B digits following the decimal point in the display.

   Since $B = 6$, the record at this stage is 0.411 764.

5. Decision: If more digits in the decimal representation are desired, continue to step 6; if not, terminate the algorithm.
6. Clear the calculator.
7. Enter a decimal point, followed by the group of B digits last recorded in step 4.

   Enter .411 764.

8. Multiply the entry in step 7 by the denominator D.

   $.411\ 764 \times 17 = 6.999\ 988$

9. Compute the difference N − (result in step 8).

   $7 - 6.999\ 988 = 0.000\ 012$

10. Convert the result in step 9 to a whole number by moving the decimal point B places to the right.

    Obtain 12.

11. Replace N in the algorithm by the whole number obtained in step 10;

The authors are on the faculty at Point Loma College in San Diego, California. Billy Hobbs is a professor in the department of mathematics and computer sciences. He is actively involved in preservice and inservice teacher education. Charles Burris teaches courses in computer science and applied mathematics and is active in developing computer applications.

then return to step 2 and repeat the same sequence of steps.

For the example 7/17, a repetition of steps 2 through 11 yields the following:

3. Note that $N = 12$ at this point, so
   $12 \div 17 = .705\ 882\ 35.$
4. The record becomes
   0.411 764 705 882.
5. Decision: Want more digits, so continue with step 6.
7. Enter .705 882.
8. Multiply by D.
   $.705\ 882 \times 17 = 11.999\ 994$
9. Compute the difference $N -$ (result in step 8).
   $12 - 9.999\ 994 = 0.000\ 006$
10. Move decimal point 6 places to the right to obtain
    6.
11. With $N = 6$, return to step 2.
3. Compute $6 \div 17$.
   $6 \div 17 = .352\ 941\ 18$
4. The record becomes
   0.411 764 705 882 352 941.
5. Decision: Once the period for 7/17 has become clear, the algorithm is terminated. As indicated earlier,
   $7/17 = 0.\overline{411\ 764\ 705\ 882\ 352\ 9}$.

Certain facts about repeating decimals may be used to make the termination of the algorithm more efficient. For example, the maximum length of the period of N/D, where $N < D$ and D has no factors of 2 or 5, is $D - 1$. Thus, in the example just used, when the record 0.411 764 705 882 352 9 with 16 digits is obtained, this must be the period (the period begins at the decimal point if D has no factors of 2 or 5). In particular, the article by Jacobs (1975) includes a table that can be used to predict the length of the period of the reduced fraction N/D, whenever the prime factors of D are less than 1000.

The algorithm is based on the operations in the procedure for long division. The only difference lies in the number of digits that are added to the quotient at each division. In the pencil-and-paper method, one digit at a time is added.

Example:

```
      .4
17 )7.0
    6 8
      2
```

This is equivalent to the algorithm in the case when $B = 1$. The next step would be to divide 17 into 20 and thus calculate the second digit.

In the algorithm for a minicalculator, a block of B digits is added to the quotient at each pass through steps 2 through 4. The exact remainder is then calculated in steps 6 through 11.

Example:

```
      .411 764  ←steps 4, 7
17 )7.000 000
    6.999 988  ←step 8
           12  ←step 10
```

The steps of the algorithm are related to the long division process by the arrows. At this point, the division can continue; however, the new numerator is 12.

Once the algorithm is understood, the record may be modified to include the current value of N written under the symbol $\wedge$. For example, a sequence of such records for the example 7/17 is shown in figure 1. This same example (7/17), done on a calculator with #(display) = 7, so that $B = 7 - 2 = 5$, yields a record as follows:

$$0_{\wedge 7}.411\ 76_{\wedge 8}4\ 705\ 8_{\wedge 14}82\ 352\ 9_{\wedge 16}41\ 17$$

The entire sequence of steps 2 through 11 is done four times; then the steps 2 through 4 are done, terminating at step 5.

Fig. 1

Record	Step
$0\,._{\wedge 7}$	3
$0\,._{\wedge 7}\,411\ 764$	4
$0\,._{\wedge 7}\,411\ 764\,_{\wedge 12}$	3
$0\,._{\wedge 7}\,411\ 764\,_{\wedge 12}\,705\ 882$	4
$0\,._{\wedge 7}\,411\ 764\,_{\wedge 12}\,705\ 882\,_{\wedge 6}$	3
$0\,._{\wedge 7}\,411\ 764\,_{\wedge 12}\,705\ 882\,_{\wedge 6}\,352\ 941$	4

The algorithm is applied to two additional examples:

Example: 33/7, with #(display) = 8. Then $B = 8 - 1 = 7$.

1. Initial record, 4. Then use $N = 5$.
3. $5 \div 7 = .714\ 285\ 71$
4. New record: 4.714 285
5. Since the period of 33/7 is at most 6, step 4 yields the period
   $$\frac{33}{7} = 4.\overline{714\ 285},$$
   and the algorithm is terminated.

Example: 13/19, with #(display) = 8
Then $B = 8 - 2 = 6$.

1. Initial record, 0.
3. $13 \div 19 = .684\ 210\ 53$
4. New record: 0.684 210
5. Decision: More digits are required; proceed to step 6.
7. Enter .684 210.
8. $.684\ 210 \times 19 = 12.999\ 99$
9. $13 - 12.999\ 99 = .000\ 01$
10. Move decimal point 6 places right; obtain 10.
11. Set $N = 10$. Return to step 2.
3. $10 \div 19 = .526\ 315\ 79$
4. New record; 0.684 210 526 315
5. Decision: More digits are required; proceed to step 6.
7. Enter .526 315.
8. $.526\ 315 \times 19 = 9.999\ 985$
9. $10 - 9.999\ 985 = .000\ 015$
10. Move decimal point 6 places right to obtain 15.
11. Set $N = 15$. Return to step 2.
3. $15 \div 19 = .789\ 473\ 68$
4. New record:
   0.684 210 526 315 789 473
5. Decision: Since the period of 13/19 is at most 18, the last record in 4 yields
   $$\frac{13}{19} = 0.\overline{684\ 210\ 526\ 315\ 789\ 473}$$
   and the algorithm is terminated.

The following decimal representations may be used to practice the algorithm:

$$\frac{62}{7} = 8.\overline{857\ 142}$$

$$\frac{50}{13} = 3.\overline{846\ 153}$$

$$\frac{1}{17} = 0.\overline{058\ 823\ 529\ 411\ 764\ 7}$$

$$\frac{39}{19} = 2.\overline{052\ 631\ 578\ 947\ 368\ 421}$$

$$\frac{19}{84} = 0.22\overline{6\ 190\ 47}$$

$$\frac{47}{65} = 0.7\overline{23\ 076\ 9}$$

$$\frac{24}{23} = 1.\overline{043\ 478\ 260\ 869\ 565\ 217\ 391\ 3}$$

$$\frac{125}{31} = 4.\overline{032\ 258\ 064\ 516\ 129}$$

$$\frac{117}{328} = 0.356\ \overline{707\ 31}$$

$$\frac{1}{29} = 0.\overline{034\ 482\ 758\ 620\ 689\ 655\text{-}}\ \overline{172\ 413\ 793\ 1}$$

References

Jacobs, Neal. "More on Repeating Decimals." *Mathematics Teacher* 68 (March 1975): 249–52.

Prielipp, Robert W. "Decimals." *Arithmetic Teacher* 23 (April 1976): 285–88.□

A Calculator Activity That Teaches Mathematics

By **Glenda Lappan** *and* **Mary Jean Winter**

The purpose of this article is to present a "calculator activity" that has many variations and affords many opportunities for the teaching of mathematics to children in the upper-elementary grades. It has been received with enthusiasm by children in grades three through five.

The Game

The activity is a variation of bingo. The basic format is a square of thirty-six numbers and under the square is a list of nine factors. (Fig. 1) The instructions for the basic game state that each number in the grid is the product of two of the numbers in the factor list. Players are to select two numbers from the list of factors, compute the product, locate the product on the grid, and write the factors in the appropriate product square. For example, suppose a student selected the factors 59 and 19 from the list. The student would calculate the product of 59 × 19, find the square containing the product 1121, and write the factors in the square.

1121
59 × 19

Each of the factors may be used more than once. No product has a repeated factor.

Fig. 1

Bingo

368	91	1652	299	161	437
364	304	301	2537	644	592
688	247	2183	532	767	112
1591	1121	703	133	1204	208
259	1357	413	1036	481	944
817	559	851	448	196	989

Each number on the grid is the product of two of the numbers below.

7 23 43 59 16 13 37 19 28

When you find a product on the grid, write in the numbers that make it. Try for a bingo.

Playing the Game

The game can be an "individual activity"—solitaire—and can be timed or untimed. Or several children may play against one another, competing to be the one to first match factors to six product squares in a row (a bingo). A group of three or four children could also work cooperatively. In the last case, the interaction among the children can lead them to make better selections.

Since every choice of two factors leads to a product on the grid, each student feels a sense of accomplishment as he writes in more and more squares. In the classes we have worked with, students have wanted to keep going until a bingo is found, even though others in the room had long since finished. In one fifth-grade class, two students devised an adaptation of tic-tac-toe. They took turns selecting factors; each labeled the products he found with his initials (as well as the factors). In the excitement of the game, neither child tried to block the other's potential bingo.

We have found that most children make their first several selections of two factors at random (at least with no discernible mathematical reason for the selection). Third graders tend to con-

Both authors are professors of mathematics at Michigan State University. Glenda Lappan works with preservice and inservice elementary and secondary school teachers and with elementary school children. She has a special interest in the use of calculators and computers in the teaching of mathematics. Mary Jean Winter is a numerical analyst by training, but she has become involved with the use of calculators and computers in the teaching and learning of mathematics. Both regularly work with children in the fourth and fifth grades, using calculators in classroom activities with the children.

tinue selecting pairs of numbers in this way, but older children, once two or three squares in the same row or column have been filled in, will concentrate on trying to complete the bingo. Comments from us like the following helped students to make better selections of factors:

What square are you trying to get?

I see. Which two numbers did you try last?

Was your answer too big or too small?

What can you try that will give you a bigger (smaller) answer?

Some children applied the ideas of evenness and oddness. If the desired product was odd, most fourth- and fifth-grade children, when asked about the factor 16, promptly said, "Sixteen can't be one because it's even," although they had not initially considered evenness or oddness when they selected factors.

Only once in our experience did a student independently figure out that to get the grid square he needed, it was sufficient to attempt to divide the entry in that square by each factor until a decimal-free answer appeared on his calculator. We were able to lead other students to this strategy by a series of questions.

From time to time we encouraged or guided students with questions like the following:

I see that the number you want ends in a 9. Does 43 times 59 end in a 9?

Can you think of a way that you can be sure you do *all* possible ways of taking two numbers from the list?

How many different bingos can you get on this board?

Several children insisted on finding all fourteen bingos. A discussion of an exhaustive strategy would have saved many an "Oh! We already have that one."

An activity related to the development of an exhaustive strategy would be to supply a list of nine numbers and ask the children to make up a bingo board for the next class to solve. The need for a systematic way of forming products would be hard to ignore.

Extending the Activity

For an older group of students you could ask questions related to the number of possible products, for a given number of factors, and ways in which the products could be arranged in the square grid.

How many products would ten distinct factors generate?

What is the smallest square grid you can fit these products into?

If the rest of the grid is filled with numbers that are impossible to get with the given factors, can you place these "impossibles" so that no bingo is possible?

Answers to questions like these can involve some sound mathematics and some worthwhile problem-solving experiences. The mathematics can be on a sophisticated level. For example, the number of possible products from ten factors can be determined by a formula for the combinations of n things taken r at a time.

$$C(n, r) = \frac{n!}{r!(n-r)!}$$

Substituting 10 for n and 2 for r, we have the following:

$$\begin{aligned} C(10, 2) &= \frac{10!}{2!(10-2)!} \\ &= \frac{10 \times 9}{2} \\ &= 45 \end{aligned}$$

But reference to the formula is not necessary. The answer to the question, How many products would ten distinct factors generate? can be determined by systematic trial.

Once we have the products, how are they to be arranged in the grid? If we can get 45 distinct products, these products will fill all but 4 squares in a 7-by-7 grid. A 7-by-7 square has 16 possible bingos, if all entries in the squares are possible. But not all entries are possible; we have 4 "impossibles." The 4 "impossibles" can be placed so as to make as many as 10 bingos or as few as 4 bingos. The result can be an easy or a tricky bingo board.

For 8 factors, however, there are 28 products. In a 6-by-6 grid, the 8 impossible answers can be placed so that no bingos can be found. It is possible to place the 8 "impossibles" so that 8 bingos remain.

There is also mathematics on the level of the elementary grades. If a certain product is desired, the selection of a likely pair of factors involves estimation, and estimation requires an understanding of number properties and the idea of rounding numbers. Suppose a girl who wants to find the factors of 703 tries 23×37. She observes that 851 is too big. If she can say to herself, "I'll try 19×37. That will be a little smaller," she is showing an understanding of one of the basic properties of inequalities:

If $a > b$ and $c > 0$, then $ac > bc$.

In this case, $a = 23$, $b = 19$, and $c = 37$. By trying 19×37 instead of 13×37, she is showing perception of the stability of multiplication; a small change in one factor produces a relatively small change in the product. She is using mathematics, in other words.

One of the basic techniques in problem solving is answering the question, What do we know cannot be the answer? As an example, consider the selection of likely factors for 1121. An immediate observation is that 1121 is an odd number—an odd number cannot have an even factor. Therefore, neither 16 nor 28 is a possible factor. Another observation is that 1121 ends in 1; the only products you can get from the factor list that will end in 1 are two factors ending in 9 (because $9 \times 9 = 81$) or one factor ending in 3 and the other in 7 (because $3 \times 7 = 21$). Therefore the possible factorizations have been restricted to the following:

23×7, 43×7, 13×7, 23×37,

43×37, 13×37, and 59×19.

Estimation and rounding will eliminate some of these candidates. The product of a two-digit number times a one-digit number is less than or equal to the biggest two-digit number times the biggest one-digit number; that is, 99×9. In other words, we know that the product of a two-digit number and a one-digit number has no more than three digits. So, 23×7, 43×7, and 13×7 will be too small. If we look individually at the others, we can see that $13 \times 37 < 20 \times 40$, and we know that $20 \times 40 = 800$, which is too small, and that $43 \times 37 > 40 \times 30$, and $40 \times 30 = 1200$, which is too big. There remain only two more pairs of factors to try. (On a more sophisticated level, we could say that we have made repeated use of the fact that if $a < b$, $c < d$, and $c \geq 0$, then $ac \leq bd$.)

The idea of finding a factor of 1121 by dividing 1121 by several numbers requires an awareness of the relationship between multiplication and division. Mathematically, we say that the operations are inverse to one another. In practice, a boy may say to himself, "If 23 is a factor of 1121, then 23 divides 1121." This awareness of the inverseness of multiplication and division is not immediate; it comes with experience.

The making of a bingo grid is a mathematical experience. The development of a method of obtaining all possible products from the list of factors, without repeating any work already done, requires an exhaustive strategy. To take 7 times all the other numbers, then 23 times all the others, then 43 times all the others, and so on, will produce all the products, but the students will soon realize that each product is being found twice. What must be devised, of course, is a listing of all combinations of 2 things (factors in this case) taken from a set of 9 things.

Postscript

The game of bingo, as described so far, is a calculator activity. The emphasis is on problem solving, and speedy calculations are needed. A calculator-free version of bingo makes an entertaining enrichment (or a disguised drill and practice) activity for younger children who are learning the multiplication facts. Figure 2 is an example. Any combination of six one-digit factors will do.

Fig. 2

10	32	24	12
56	15	28	6
20	FREE	14	35
21	8	40	16

Each number on the grid is the product of two of these numbers

2 3 4 5 7 8.

You could focus on the "hard" combinations by using the factors 4, 5, 6, 7, 8, and 9.□

Using a Calculator to Do Arithmetic in Bases other than Ten

By **James E. Schultz**

The method described here for doing arithmetic in bases other than ten on an ordinary calculator is of interest not only because of its direct application to computation, but also, perhaps more so, because of the example it provides of the folly of jumping to conclusions too quickly. In the hands of an effective teacher, this method can be used both to foster a better understanding of place value and to illustrate the danger of blind dependence on the calculator.

We begin by describing the computational procedure and conclude by discussing the limitations. Since the ordinary calculator is limited to base-ten arithmetic, we will convert a given problem in another base to a related problem in base ten, solve the related problem in base ten, and convert the answer back to the original base. The technique described below can be done quickly on most calculators and does not require a memory feature. To fully realize the ease with which the computations can be carried out, we recommend that the reader use a hand calculator to work through the examples as they are presented.

James E. Schultz is an associate professor in the Department of Mathematics at The Ohio State University in Columbus. He specializes in teacher training programs at Ohio State and has also been involved with programs at Indiana University, Michigan State University, and the University of Chicago.

The underlying idea is to use a block of two (or more) digits on the calculator to represent each of the characters in the given base. For example, the calculator representation for 234_{five} using blocks of two digits is 20304. The last two digits (04) indicate that we have 4 ones, the 03 indicates we have 3 fives, and the 2 indicates that we have 2 twenty-fives. Zeroes are inserted to establish blocks of two digits for each of the places of the 234_{five}. (The need for blocks of two or more digits will become apparent shortly.) Some additional examples are given in figure 1, where T = ten and E = eleven.

The problem $13_{five} + 21_{five}$ is then transformed (using a block size of two) as follows:

$$\text{from}\quad \begin{array}{r} 13_{five} \\ +\,21_{five} \\ \hline \end{array} \quad \text{to} \quad \begin{array}{r} 103 \\ +\,201 \\ \hline \end{array}$$

Solving the problem (on the calculator) gives 304, which is recognized to be (in base five) 34_{five}, the correct answer. In future problems we will write simply as follows:

$$\begin{array}{rcr} 13_{five} & & 103 \\ +\,21_{five} & \rightarrow & +\,201 \\ \hline 34_{five} & \leftarrow & 304 \end{array}$$

Next we consider a problem involving regrouping.

$$\begin{array}{rcr} 24_{five} & & 204 \\ +\,13_{five} & \rightarrow & +\,103 \\ \hline ? & \leftarrow & 307 \end{array}$$

Since we can regroup seven into one five and two ones, seven is not a permissible base-five character. Together with the three fives we already had, this yields a total of 4 fives and 2 ones, which is written 42_{five} or in our calculator notation 402. We can summarize this by saying that on seeing a character greater than four (the largest permissible single character in base five) we must decrease this block by five and increase the next block to the left by one. This trading of five ones for one five in terms of our calculator notation (assuming two-digit blocks) means subtracting 5 and adding 100, which has a net effect of adding 95. So our computation can be shown by the following:

$$\begin{array}{rcr} 24_{five} & & 204 \\ +\,13_{five} & \rightarrow & +\,103 \\ \hline & & 307 \\ & & +\,95 \\ \hline 42_{five} & \leftarrow & 402 \end{array}$$

The same idea works in exchanging in other positions as illustrated in the next example, where fives are exchanged for twenty-fives.

$$\begin{array}{rcr} 21_{five} & & 201 \\ +43_{five} & \rightarrow & +403 \\ \hline & & 604 \\ & & +9500 \\ \hline 114_{five} & \leftarrow & 10104 \end{array}$$

Other bases are done in the same way. To exchange ones for sixes in base six we would subtract six and add 100 (namely, add 94).

$$\begin{array}{rcr} 24_{six} & & 204 \\ +13_{six} & \rightarrow & +103 \\ \hline & & 307 \\ & & +\,94 \\ \hline 41_{six} & \leftarrow & 401 \end{array}$$

Similar exchanges are readily developed for other bases.

The need for blocks of two or more (even when the base is less than ten) is shown in the next example.

$$\begin{array}{rcr} 47_{eight} & & 407 \\ +64_{eight} & \rightarrow & +604 \\ \hline & & 1011 \end{array}$$

Before regrouping we have 10 eights and 11 ones, more than can be displayed on the calculator in blocks of a single digit. With regrouping we have

$$\begin{array}{rrl} & 1011 & \\ & +\ \ 92 & \text{(exchange 8 ones for 1 eight)} \\ & +9200 & \text{(exchange 8 eights for 1 sixty-four)} \\ 133_{eight} \leftarrow & 10303 & \end{array}$$

Only a minor modification is necessary to adapt this method to subtraction. We leave the discovery of the technique for doing subtraction to the reader, with the hint that the necessary modification is appropriately based on the idea that subtraction is the inverse of addition.

The procedure for multiplication is analogous. One example is discussed in some detail; three others are sketched. In the example in figure 2, we transform the problem as in previous examples and multiply using the calculator. When the product 21320 is obtained, it becomes necessary to initiate exchanges, beginning with the 20 in the ones block. After two exchanges (adding 92 twice) this block becomes 04, in range for base-eight notation. We then look at the next two-digit block (15) and exchange by adding 9200 to get 07. Finally, we transform back to base-eight notation. We will subsequently abbreviate this to the following:

$$\begin{array}{rcr} 25_{eight} & & 205 \\ \times\,14_{eight} & \rightarrow & \times\,104 \\ \hline & & 21320 \\ & & +\ \ 92 \\ & & +\ \ 92 \\ & & +9200 \\ \hline 374_{eight} & \leftarrow & 30704 \end{array}$$

Fig. 1

Number	Calculator representation	Block size
13_{five}	103	2
114_{five}	10104	2
7108_{nine}	7010008	2
43_{twelve}	403	2
$E5_{twelve}$	1105	2
92_{twelve}	9002	3
$34T_{fifteen}$	3004010	3

In practice we would observe the respective blocks of the calculator display to determine when a sufficient number of exchanges had occurred; namely, when the numbers in the blocks fall into the range of the base involved.

Two more examples follow:

43_{twelve}		403
$\times\ T5_{twelve}$	→	× 1005
		405015
		+88
		+8800
		+8800
		+8800
		+8800
		+880000
		+880000
		+880000
3833_{twelve}	←	3080303

282_{nine}		20802
$\times\ 24_{nine}$	→	× 204
		4243608
		+9100
		+9100
		+9100
		+9100
		+910000
		+910000
		+910000
7108_{nine}	←	7010008

The last example suggests that the exchanging be done from right to left, as blocks to the right affect blocks to the left. (Here "carries" from the 36 affected the 24.)

Fig. 2

25_{eight}		205	(transform to calculator notation)
$\times\ 14_{eight}$	➡	× 104	(transform to calculator notation)
		21320	(multiply using calculator)
		+ 92	(exchange 8 ones for 1 eight)
		21412	
		+ 92	(exchange 8 ones for 1 eight)
		21504	
		+9200	(exchange 8 eights for 1 sixty-four)
374_{eight}	⬅	30704	(transform to base eight notation)

The next example demonstrates the adaptability of this method even to larger bases.

$E3_{nineteen}$		1103
$\times\ 72_{nineteen}$	→	× 702
		774306
		+8100
		+8100
		+810000
		+810000
		+810000
		+810000
$4356_{nineteen}$	←	4030506

One limitation of this computational technique becomes readily apparent when doing multiplication. It involves the number of digits displayed on the calculator. Since three-digit numbers require five (or six) digits in their calculator representation, the product of two such numbers would already require at least nine digits in its calculator representation. Thus this process is capable of handling only relatively small numbers, especially when the operation is multiplication.

A second limitation, which may escape the less cautious student, results from the fact that a two-digit block is not always sufficient to accommodate the numbers that arise, particularly when doing multiplication. Consider for instance the following computation:

E_{twelve}		11
$\times\ T_{twelve}$	→	× 10
		110

In this problem the product of two one-digit base twelve numbers requires *three* digits in its calculator representation. The representation of numbers in two-digit blocks would suggest an answer of $1T_{twelve}$, which is wrong. However, the problem can be done using three-digit blocks as follows:

E_{twelve}		11	
$\times\ T_{twelve}$	→	× 10	
		110	
		+988	nine times
		+988	
		+988	
		...	
		+988	
92_{twelve}	←	9002	

Unfortunately, using three-digit blocks serves to intensify the first limitation referred to earlier.

It should be noted that the second limitation can arise even in bases less than ten as shown in the next problem.

77_{eight}		707
$\times\ 77_{eight}$	→	× 707
		499849
		+92
		+92
		+92
		+92
		+92
		+92
		500401 ...

Note that the contents of the block marked 04 "spilled over" into the next block (to the left), rendering the solution invalid. (Carrying out the balance of the computation by the procedure described yields an incorrect solution of 6241_{eight}.) The correct solution can be obtained using three-digit blocks if a calculator with ten-digit display or with scientific notation is available. (In the example shown, it turns out that trading eights before trading ones would also alleviate the problem.)

The method that has been described is dependable for adding or subtracting pairs of numbers within the display limitation of the calculator being used, but the limitations certainly obviate rote application to multiplying numbers. When exploited correctly by the teacher, there may be more value derived from the cases where the procedure fails than those where it works.□

ACTIVITIES

Edited by
L. CAREY BOLSTER and EVAN M. MALETSKY
Baltimore County Public Schools Towson, Maryland
Montclair State College Upper Montclair, New Jersey

CAN YOU PREDICT THE REPETEND?

By Douglas Woodburn, Perry Hall Junior High School, Baltimore County, Maryland

Teacher's Guide

Grade level: 7–12

Materials: Calculator; activity sheets for each student

Objective: Students will use a calculator to discover patterns that will allow them to predict the repetend when a number is divided by 9, 99, and 999.

Discussion: This activity should not be attempted unless students have access to a hand calculator or a table-top model. For many students, the term *repetend* may be new. Be sure they understand that the repetend is the repeating digit or series of digits in a repeating decimal.

The general pattern for predicting the repetend is this:

1. The number of 9s in the divisor is the number of digits in the repetend. If dividing by 9, the repetend will be one digit; by 99, two digits; by 999, three digits; and so on.

2. Dividing by 9
   Separate the digits and add. If the sum has more digits than the repetend, add the digits in the sum.

   5885 ÷ 9
   5 + 8 + 8 + 5 = 26
   2 + 6 = 8
   The repetend is 8.

3. Dividing by 99
   Group digits in pairs from right to left. Then use the same process as when dividing by 9.

   2689 ÷ 99
   26 + 89 = 115
   1 + 15 = 16
   The repetend is 16.

4. Dividing by 999
   Group digits in threes from right to left, and use the same process.

   13562 ÷ 999
   13 + 562 = 575
   The repetend is 575.

These patterns should develop as the students work through sheets one, two, and three. It is suggested that the teacher do the activities prior to using them with students in order to anticipate students' questions.

You may wish to extend the activity by finding the pattern when dividing by 11.

Can You Predict the Repetend?

Divide by 9.

The repetend is the repeating digit or series of digits in a repeating decimal. $123 \div 9 = 13.66\overline{6}$. The repetend is 6. $728 \div 99 = 7.35\overline{35}$. The repetend is 35. In this activity, you will be observing patterns and developing rules for predicting the repetend.

1. Use a calculator for all divisions. Complete the chart.

DIVIDEND	4	6	3	2	1	7	8	5
Divide by 9								
REPETEND	4							

2. Write the pattern you observe for dividing a single digit by 9.

3. Find the repetend for these numbers:

DIVIDEND	12	16	32	43	25	60	71	115	204	143
Divide by 9										
REPETEND										

4. Write the pattern for finding the repetend. (HINT: Look at the sum of the digits in the dividend.)

5. Find the repetend.

DIVIDEND	58	86	38	46	98	139	433	168	5613	48167
Divide by 9										
REPETEND										

Look at your results in problem 5. Since the divisor (9) has only one digit, the repetend will have only one digit. Find the sum of the digits in the dividend, and then add the digits in the sum until the sum is a single digit.

6. Use the rule stated above to predict the repetend. Then check your predictions with the calculator.

DIVIDEND	73	435	6245	35265	42356	88888	52736	982645
Divide by 9								
PREDICTED REPETEND								
REPETEND								

Can You Predict the Repetend?

Divide by 99.

You developed a rule for predicting repetends when dividing by 9. See if it applies to division by 99.

1. Complete the chart. Use a calculator.

DIVIDEND	6	8	23	46	89	64	73	37	49	87	44	88
Divide by 99												
REPETEND												

2. What pattern do you observe?

____

3. Find the repetend.

DIVIDEND	125	421	142	283	450	804	716	826	489
Divide by 99									
REPETEND									

4. Write the pattern for finding the repetend. (HINT: Separate 125 into 1 and 25.)

____

5. Since the divisor 99 has only two digits, the repetend will have only two digits. Predict the repetend, then check with the calculator.

DIVIDEND	5	23	246	745	289	587	868	598	2315	8549	76834
Divide by 99											
PREDICTED REPETEND											
REPETEND											

Can You Predict the Repetend?

Divide by 999.

1. Complete the chart.

DIVIDEND	3	8	25	63	89	462	123	809	900
Divide by 999									
REPETEND									

A pattern seems obvious, but let's try some larger numbers.

2. Complete the charts.

DIVIDEND	1402	3512	4132	5627	4839	8786	8569	7348
Divide by 999								
REPETEND								

DIVIDEND	31124	25631	79876	85697	401132	23167	112678
Divide by 999							
REPETEND							

3. Describe the pattern for finding the repetend when a number is divided by 999.

____

BONUS:

4. A rule <u>does</u> exist for predicting repetends when dividing <u>any</u> number by <u>any</u> member of the set 9, 99, 999, 9999, and so forth. Try to develop it.

Calculator ideas for the junior high classroom

EARL OCKENGA

Currently a classroom teacher at the Malcolm Price Laboratory School, University of Northern Iowa, Earl Ockenga is on leave from Axtell Park Junior High School, Sioux Falls, South Dakota.

One of my seventh graders wagered, as he handed me his calculator, "Betcha don't know how many times my heart has beat in the last year." "About 36 792 000 times," he said in response to my surprised look. "This newspaper ad isn't correct," announced another student as she showed me her calculator. "The ad says 20% off on all items, but it's really only 16.7% off on the dishes." Comments such as these occur when hand-held calculators are made available in the classroom.

There is no doubt that junior high students like to use calculators. It's fun, for example, to first imagine a stretch of highway reaching to the sun, and then, with the aid of a calculator, to compute to find that a trip to the sun, at the speed of 95 kph, would take 193 years. Being able to compute accurately is a good feeling, and calculators do make it possible for students to compute quickly and precisely. But, calculators can be more than just machines used to get right answers. These amazing little devices can also become powerful instructional tools in the classroom.

COMPUTATION AND ESTIMATION SKILLS

Getting students to brush up on their basic facts is not always easy. However, with a calculator and a game called Check It, the task can become easier. Here's how my students play the game:

- The person holding the calculator calls out a basic fact combination such as (9 × 6) + 11, and then enters 9 ⊠ 6 ⊞ 11 in the calculator.
- The other players write down their answers. Then the caller pushes the ⊟ key and gives the answer.
- Players having the correct answer score 1 point. The calculator is then passed to another player who calls out a new fact combination.

The game of Check It is an enjoyable way to practice basic facts and it's intriguing how the player who is doing the calling seems to know the basic fact weaknesses of the other players.

There are many ways the calculator can be used to help students develop their estimation skills. A favorite of mine is this game for two or more players. All that's needed is one calculator and a set of division exercises.

72 345 ÷ 23	10 469 ÷ 63
14 503 ÷ 603	71 725 ÷ 909
18 861 ÷ 7123	90 863 ÷ 14

- The game starts by selecting one of the division exercises.
- Each player writes down his or her estimate of the answer.
- The calculator is used to determine the exact answer.
- Players score 1 point if their estimate has the correct number of digits. Players score 2 points if their estimate has both correct number of digits and correct first digit.
- Play continues until a predetermined score is reached.

The calculator can also be used to encourage the development of mental computational skills. When doing exercises like the following on the calculator, most students soon discover that when multiplying by 10s and 100s, "using your head" is faster than using the machine.

37 × 10____	37 × 100 ____
123 × 10____	123 × 100____
3.5 × 10____	3.5 × 100____
.74 × 10____	.74 × 100____

Sequences of related exercises are often an effective means to get at hard-to-teach ideas. The following activity can help students who are having trouble with the rules for placement of decimal points in division.

Try these on a calculator.

.0125 ÷ .005 = ____
.125 ÷ .05 = ____
1.5 ÷ .5 = ____
12.5 ÷ 5 = ____

Why are all the answers the same?

The follow-up is to have students write their own set of related division problems. Students can now confirm their thinking by using a calculator to check their exercises. It's surprising how often the calculator can help accomplish what no amount of teacher telling-and-showing can do.

Write a division exercise for each answer.
Use only the digits 0, 4, 7, and 8.

.0084 ÷ .007 = 1.2
________ = 1.2
________ = 1.2
________ = 1.2

MEASUREMENT AND GEOMETRY

Metric measure and the calculator are a natural combination for getting students involved in measurement activities. With metric tapes students can measure the circumference and diameter of tin cans and then compute an approximation for pi. Using the calculator they can perform this activity for cylinders of many different sizes, thereby making the generalization of how circumference is related to diameter more obvious.

Having students find the area of at least one triangle by measuring all three bases and all three heights is another example. With the aid of a calculator the focus of the activity is not on computation, but rather on the meaning of $A = .5bh$.

Four or five cardboard boxes of different shapes, metric tapes, and a calculator as a resource, can become the starting point for measurement activities like these.

1. Make a contest out of estimating the volumes of the boxes. Take the measurements and check the estimates by computing with the calculator.
2. Find the box that has a surface area-to-volume ratio closest to 1.
3. Use a calculator to help decide how to cut down a box to a volume of 1000 cubic centimeters. (Your class could end up with an interesting collection of liter containers of different shapes.)
4. Turn a box bottom up, and decide where to make cuts to get a cardboard pyramid that has a surface area of 1500 square centimeters.

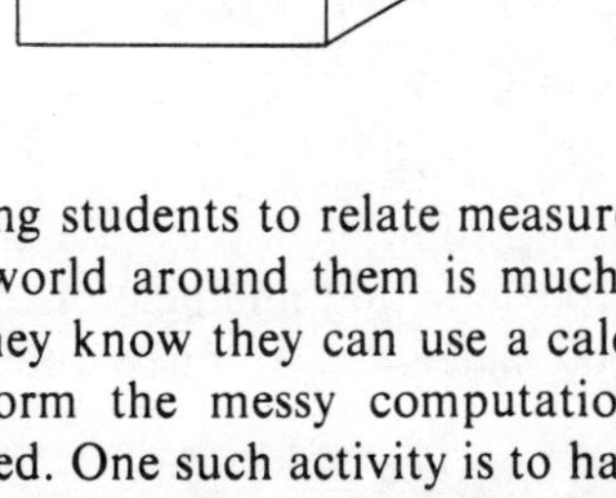

Getting students to relate measurements to the world around them is much easier when they know they can use a calculator to perform the messy computations encountered. One such activity is to have students make measurements and then describe how much is needed to make a million. The following are examples:

- 8 196 721 cans of Shasta Draft Root Beer would stack to a height of 1 000 000 meters.

- 19 469 people just my size would weigh a total of 1 000 000 kilograms.
- It takes 1 587 301.5 footsteps for me to walk 1 000 000 meters.

A bulletin board collection of "make a million" facts would keep some students measuring and computing long after the formal classroom study of metric measure has been completed.

FUNCTIONS

Introducing students to the idea of a function can be greatly enhanced by means of the calculator. Most students are familiar with the "machine" idea of a function as seen in textbook pictures, and they have little difficulty in recognizing that their calculator could act as an electronic function machine.

The idea of using the calculator as a function machine occurred the day my students were establishing the length of their pace. One student's pace was .6 meter. To show how distances could be estimated by knowing the length of the pace, I wrote

$$p \boxed{\times} .6 \boxed{=} d$$

on the chalkboard. It was surprising how easily they could enter a number in their machine and then push

$$\boxed{\times} .6 \boxed{=}$$

to determine distance. In no time at all we had collected a set of input-output numbers.

Number of paces (p)	75	4565	24 300	17.5
Distance in meters (d)	45	2739	14 580	10.5

My suggestion that they use their machine to find the number of paces in 1 000 000 meters led to a new code,

$$d \boxed{\div} .6 \boxed{=} p,$$

which we agreed to write as

$$\frac{d}{.6} = p.$$

Others in the class wanted to write a function code for their pace, and before long we were investigating what advantage a taller person might have over a shorter person. To better see the individual pacing comparisons, we graphed

$$\frac{d}{.6} = p$$

and

$$\frac{d}{.75} = p.$$

Everyone helped by using their calculators to generate sets of ordered pairs. Some students were surprised to see that regardless of their choice of input number, the ordered pairs resulted in points on the same line.

When using a calculator, it's fun to make up your own function codes such as these to investigate and graph.

$$x \boxed{+} 3 \boxed{=} y$$
$$x \boxed{-} 3 \boxed{\times} 7 \boxed{=} y$$
$$x \boxed{\div} 3 \boxed{-} 4 \boxed{=} y$$
$$x \boxed{\times} \boxed{=} \boxed{+} 3 \boxed{=} y$$

With a little experience, most students become quite good at rewriting $x \boxed{\times} \boxed{=} \boxed{+} 3 = y$ in the form $x^2 + 3 = y$.

A calculator can also be helpful in making interpretations of data. Once, as a class project, we weighed various lengths of wire and recorded our findings on a table. Then, with calculators, students analyzed the data entries by looking for a constant ratio.

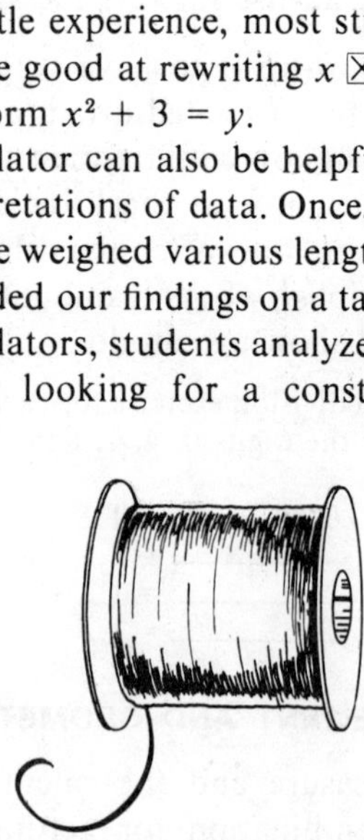

centimeters of wire (cm)			
75	100	150	180
grams of wire (g)			
4.9	6.4	9.7	11.9

Translating what they did on their calculators into a rule such as

$$c \boxed{\times} .065 \boxed{=} g$$

or

$$g \boxed{\div} .065 \boxed{=} c$$

was an easy step for most students.

PROBLEM SOLVING AND APPLICATIONS

It came as a shock that story problems, which usually took 50 minutes to complete with paper and pencil, could now be completed in less than 15 minutes using the calculator. It was apparent that new problem-solving situations were needed. Newspapers and almanacs became idea sources for calculator problem settings.

Students given these 1975 baseball statistics, for example, used their calculators to compute the batting average, home run average, and runs-batted-in average of each player. Students then compared the average to decide who was the most "dangerous" hitter. After careful comparisons and a few arguments, most, but not all, agreed Scott was the most dangerous hitter.

Player	*AB*	*H*	*HR*	*RBI*
Schmidt	562	140	38	95
Jackson	593	150	36	104
Scott	617	176	36	109
Kingman	502	116	36	88
Bench	530	150	28	110

All sorts of real-world settings can become good calculator problem-solving situations. Teams of four students, for example, were asked to determine the number of people needed to overload one of our local department store elevators. The only information they were given was that which appeared on this sign found on the wall of the elevator.

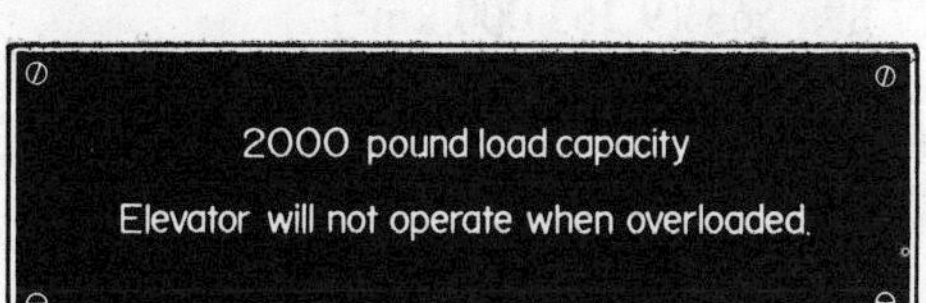

After collecting data on the weights of people, making estimates, and performing dozens of messy computations, the students decided that either 12 adult men, 16 adult women, 19 seventh graders, or 8 NFL linemen would more than likely overload the elevator. Because all members of the teams could, when using a calculator, perform calculations quickly and accurately, the primary task of the activity was not tedious computations, but rather the problem-solving process.

Even the calculator itself can become the setting for problems. Try this on your calculator. Using only the 3, $\boxed{+}$, $\boxed{-}$, $\boxed{\times}$, $\boxed{\div}$, and $\boxed{=}$ keys, see if you can make your machine display all the whole numbers from 1 to 10.

SUMMARY

Many other topics in the junior high grades such as exponents, percents, solving equations, and even common fractions are good areas for exploration in effective uses of the calculator. The nice part is you don't need a classroom set of calculators to begin the exploration. Even with a few machines you and your students can experience ways in which calculators can make the teaching task easier and the opportunities for learning more enjoyable.

CALCULATIONS YOU WOULD NEVER MAKE WITHOUT A MINICALCULATOR

An excellent collection of interesting problems for minicalculators.

By BONNIE H. LITWILLER
and DAVID R. DUNCAN

University of Northern Iowa
Cedar Falls, IA 50613

Students of all ages are intrigued by large numbers. They are even more interested when these numbers are the results of computations with real data. These problems may be serious applications or may simply result from someone's curiosity.

All of our suggested classroom activities are more curious than serious and would have been most difficult without the use of a minicalculator. The data sources were encyclopedias and almanacs, since these are readily available in media centers.

Activity I

The Grand Canyon is an immense natural phenomenon whose length is 350 km. Its width varies from 6.5 km to 29 km, and its maximum depth is 1737 m. Using 350 km as its length, 16 km as its average width, and 914 m or 0.914 km as its average depth, the volume of the Grand Canyon is approximately 5118 km³. To fully appreciate the magnitude of this number, consider the following questions:

If the Grand Canyon were filled with Arizona soil, to what depth would the soil have to be removed from the rest of the state?

Since Arizona has an area of 286 720 km² (excluding the Grand Canyon), approximately 18 meters of soil would have to be removed from the entire state to accomplish our fill project.

$$\frac{5118 \text{ km}^3}{286\,720 \text{ km}^2} \approx .018 \text{ km} = 18 \text{ m}$$

How many years would it take to fill the Grand Canyon if an amount of corn equal to the entire 1973 record corn crop were dumped in each year?

The U.S. corn production in 1973 was 19 894 300 m³ or 0.199 km³; thus, it would take approximately 25 700 years to fill the Grand Canyon with corn.

$$\frac{5118 \text{ km}^3}{0.199 \text{ km}^3} \approx 25\,700$$

Activity 2

This activity involves densities of population; the relevant data are shown in table 1.

TABLE 1

Location	Population	Area in Square Kilometers
New York City	7 895 563	828
United States	204 765 770	9 363 000
Wyoming	332 416	252 000
Bangladesh	75 000 000	143 000
World	4 000 000 000	

If the entire United States were as densely populated as New York City, what would be the population of the United States?

The population of the United States would be approximately 89 000 000 000; this is twenty-two times the world population.

$$\frac{(7\,895\,563)(9\,363\,000 \text{ km}^2)}{828 \text{ km}^2} \approx 89\,000\,000\,000$$

If the entire United States were as densely populated as Wyoming, what would be the population of the United States?

The population of the United States would be approximately 12 000 000.

$$\frac{(332\,416)(9\,363\,000\text{ km}^2)}{252\,000\text{ km}^2} \approx 12\,000\,000$$

If the entire United States were as densely populated as the country of Bangladesh, what would be the population of the United States?

The population of the United States would be approximately 5 000 000 000; this is more than the world population.

$$\frac{(75\,000\,000)(9\,363\,000\text{ km}^2)}{143\,000\text{ km}^2} \approx 5\,000\,000\,000$$

Activity 3

Suppose that as the current population of the world died, they were to be buried in coffins that had dimensions 2 meters by 0.5 meter by 0.3 meter.

How much space (volume) would be needed to bury these coffins if the cemetery had area 2 km²?

The population of the world could be buried in a cemetery whose length, width, and depth are 2 km by 1 km by 0.6 km.

$$\frac{(4\,000\,000\,000)(2\text{ m})(0.5\text{ m})(0.3\text{ m})}{2\,000\,000\text{ m}^2} \approx 600\text{ m} = 0.6\text{ km}$$

Activity 4

Suppose the entire population of the world came to a rock concert.

If 0.25 square meters of space (area) were allowed for each person to stand, how much space (area) would be needed for the crowd?

The space needed for the rock concert would be approximately 1000 square kilometers. This is the area of Jay County, Indiana. The entire world population could stand in one small county!

Activity 5

Suppose that, to guard against an invasion, the entire population of Iowa lined up side by side around its boundary (perimeter).

How much space (perimeter) would each person have to guard?

The perimeter of Iowa is 1545 kilometers and its population is 2 825 000; thus each person would guard approximately 0.5 meter.

$$\frac{1545\text{ km}}{2\,825\,000} \approx 0.0005\text{ km} = 0.5\text{ m}$$

The population of Iowa fits neatly about its boundary!

If the entire population of Wyoming lined up side by side to guard its state, how much space (perimeter) would each person have to guard?

Since the perimeter of Wyoming is 2125 kilometers and its population is approximately 332 000, each person would have to guard approximately 6.4 meters.

$$\frac{2125\text{ km}}{332\,000} \approx 0.0064\text{ km} = 6.4\text{ m}$$

The interested reader may wish to do this problem using a densely populated state with a small perimeter, such as New Jersey.

Activity 6

As a Bicentennial project, it was at one time proposed that a human chain be formed from coast to coast.

How many people would be required to form such a chain?

If the distance across the U.S. is 4800 kilometers and each person requires 0.8 meter, approximately 6 000 000 persons are required to form a human chain across the U.S.

$$\frac{4\,800\,000\text{ m}}{0.8\text{ m}} \approx 6\,000\,000$$

If a Bicentennial flag were passed from coast to coast by this human chain, how long would it take to pass the flag?

Allowing one-half second per person, it would require just over one month for the flag to be passed across the continent.

$$\frac{6\ 000\ 000}{2(60)(60)(24)} \approx 34.7 \text{ days}$$

(Is this faster or slower than the United States Postal Service?)

Activity 7

If all the cars in the United States were lined up bumper to bumper, how long a line would be formed?

The number of registered cars in the U.S. is 90 000 000. If the average car length from bumper to bumper is 4.6 meters, then the line formed by the cars would be 414 000 kilometers. This is 10 times around the world at the Equator. (The circumference of the Equator is 40 200 kilometers.)

$$\frac{414\ 000 \text{ km}}{40\ 200 \text{ km}} \approx 10$$

Activity 8

Seemingly small amounts of rain can yield large amounts of water.

How many metric tons of water fall during a 2.5 centimeter rain over one square kilometer?

Since water weighs 1 gram per cubic centimeter, approximately 25 000 metric tons fall.

$$\frac{(2.5 \text{ cm})(100\ 000 \text{ cm})^2}{1\ 000\ 000 \text{ grams/metric ton}} \approx \frac{25\ 000\ 000\ 000 \text{ cm}^2}{1\ 000\ 000}$$

$$\frac{25\ 000\ 000\ 000 \text{ grams}}{1\ 000\ 000} = 25\ 000 \text{ metric tons}$$

If this amount of water were to be used to fill a swimming pool whose length and width were each 25 meters, the depth of the swimming pool would have to be 40 meters.

How many metric tons of water fall during a 0.25 centimeter rain over a football field?

Since the dimensions of a football field are 360 feet by 160 feet, or 109.73 meters by 48.77 meters, approximately 13.4 metric tons of water fall.

$$\frac{(0.25 \text{ cm})(10\ 973 \text{ cm})(4877 \text{ cm})}{1\ 000\ 000 \text{ grams/metric ton}} \approx 13.4$$

Activity 9

Assuming that a bumper Iowa crop of 1 000 000 000 bushels (35 000 000 cubic meters) of corn were spread evenly across the state, how deep would the corn be?

Since the area of Iowa is 145 790 square kilometers, the entire state of Iowa would be covered to a depth of only 0.024 centimeters.

$$\frac{35\ 000\ 000 \text{ m}^3}{145\ 790\ 000\ 000 \text{ m}^2} \approx 0.00024\text{m} = 0.024 \text{ cm}$$

Activity 10

How long does it take to live 1 000 000 000 seconds?

One of the authors admits the birthdate 8 June 1940 at 11 p.m. He celebrated his 1 000 000 000 seconds on 16 February 1972 at 40 seconds after 12:46 a.m. When will you celebrate your 1 000 000 000th second? Note, you will have lived 31 2/3 years!

Interested readers and their students are invited to formulate other curious activities and perform their computations using computational devices.

CUBE ROOTS ON A CALCULATOR

By REBECCA A. EIMER

Northeast Missouri State University
Kirksville, MO 63501

SUPPOSE you are given the problem of determining the radius of a given sphere, knowing the volume to be 1910.0928 cubic units. (This figure has been selected for convenience so that we may compute the cube root of an integer.) Using $\pi = 3.1416$ and solving the equation $V = (4/3)\pi r^3$, you determine that the radius is $\sqrt[3]{456}$. But in practical situations you will need a decimal approximation for $\sqrt[3]{456}$. Of course, if you happen to have a book of tables handy, merely open it to the correct page and find the answer. But what if you do not happen to have your tables with you?

Today even grade school pupils often have an electronic calculator handy. But unless you happen to own one of the luxury models, your calculator is not capable of doing this computation—or is it? If it will figure square roots, it is; if it also has a memory, you will not even have to lift a pencil.

Just what will your calculator do? Since it gives square roots, it will, of course, also give the fourth, eighth, sixteenth roots, and so on. Looking at the sequence {1/2, 1/4, 3/8, 5/16, 11/32, 21/64, 43/128, 85/256, . . .}, you have $a_{n+1} = (1 - a_n)/2$. Taking the limit of both sides as $n \to \infty$ of $a_{n+1} = (1 - a_n)/2$, we have $L = (1 - L)/2$, and L must equal 1/3. Thus, using this sequence as exponents of 456, you can approximate $\sqrt[3]{456}$ to any desired accuracy. Each power of 456 is now of the form $456^{a/2^k}$ where a and k are integers, and your calculator does have this capability.

This is the method of computing the cube root of x, where x is a real number:

1. Enter x.
2. Take square root.
3. Take square root again.
4. Place in memory (or make a note).
5. Enter x.
6. Divide by memory.
7. Take square root.
8. Repeat steps 4 through 7 until desired accuracy is obtained.

Applying this algorithm to 456, you get the sequence of approximations in Table 1 for $\sqrt[3]{456}$. The radius of our sphere, accurate to three places, is 7.697 units.

TABLE 1
Sequence of Approximations for $\sqrt[3]{456}$

Exponent of 456	Value
1/2	21.3541565
1/4	4.6210557
3/8	9.9337177
5/16	6.7752685
11/32	8.2038774
21/64	7.4554324
43/128	7.8207067
85/256	7.6358857
171/512	7.7277437
341/1024	7.6816774
683/2048	7.7046761
1365/4096	7.6931682
2731/8192	7.6989199
5461/16384	7.6960435
10923/32768	7.6974816
21845/65536	7.6967625
43691/131072	7.6971221
87381/262144	7.6969423
174763/524288	7.6970322
349525/1048576	7.6969872
.	.
.	.
.	.

With only slight modification, you can expand the algorithm to additional roots of a given real number.

USING A CALCULATOR TO FIND RATIONAL ROOTS

Your four-function calculator can find $\sqrt[5]{50}$ . . . in fact $\sqrt[n]{m}$.

By **BERT K. WAITS**
The Ohio State University
Columbus, OH 43210

A recent article in the *Mathematics Teacher* (Eimer 1977) included a recursive method for finding the cube root of ten that used a square root algorithm. The method required only a simple four-function calculator with a square-root key and a memory. In this note we present a more direct method for approximating the *n*th root of any positive number that requires only a four-function calculator with a square-root key and repeat multiplication capability. Our method does not require a knowledge of calculus or a calculator with a memory. Most four-function calculators have the desired repeat multiplication capability as a built-in feature.

We need first to establish some preliminaries. Hereafter, we assume that a is a positive real number and that n and k are positive integers. First, note that $a^{1/2^k}$ can be computed by a repeated application of the square-root key. For example, if $k = 3$, then

$$a^{1/2^3} = a^{1/8} = \sqrt[8]{a} = \sqrt{\sqrt{\sqrt{a}}}.$$

In addition, a^{2^k} can be found by repeated squaring (or repeated multiplication, if your calculator does not have a key for squaring numbers). For example, if $k = 3$, then $a^{2^k} = a^8 = ((a^2)^2)^2$.

Our main result is stated in the following theorem. We delay its derivation until several examples have been discussed.

THEOREM: *The nth root of any positive real number a can be approximated by the number*

$$(1) \qquad b_k = \left(\frac{a^{1/2^k} - 1}{n} + 1\right)^{2^k}$$

for k sufficiently large.

Experience indicates that b_k for $10 \le k \le 15$ gives good approximations of $a^{1/n}$ (the nth root of a) for all values of $a < 10^8$ and $n < 1000$.

As an example we will approximate $50^{1/5}$, using (1) with $k = 10$.

To do this we evaluate

$$\left(\frac{50^{1/2^{10}} - 1}{5} + 1\right)^{2^{10}}$$

using only a four-function calculator with a square-root key. There is no need to record intermediate results. First enter 50 in the display, and then press the square-root key ten times. Now subtract 1, divide by 5, add 1, and then square the result ten times. Most four-function calculators without a squaring key will square repeatedly if one depresses certain keys (usually two) in the proper order. One popular and inexpensive four-function calculator yields the result that $50^{1/5} \approx 2.188\,8877$ when $k = 10$. For $k = 12$, the same calculator yields 2.185 9977, and for $k = 15$, the result we obtain is $50^{1/5} \approx 2.166\,3904$. Other four-function calculators may give slightly different results. Advanced students could be asked to investigate what happens to the approximations for larger values of k: $k = 15, 16, 17$, and so on. Compounded round-off errors account for the fact that the approximations become less accurate. For comparison, note that the actual value of $50^{1/5}$ (to eight places) is 2.186 724 15. If your students have access to a programmable calculator, then they can easily investigate the convergence of the sequence $\{b_k\}$ in (1) for different values of a and n.

We can derive our theorem from the binomial series

$$(2) \qquad (1 + \epsilon)^r = \sum_{n=0}^{\infty} \binom{r}{n} \epsilon^n,$$

where

$$\binom{r}{n} = \frac{r(r-1)(r-2)\cdots(r-n+1)}{n!}.$$

This expansion is valid for every $-1 < \epsilon < 1$ and any real number r (Apostal 1962, pp. 453–54). By writing out the first three terms of (2), we observe that

$$(3)\qquad (1+\epsilon)^r = \binom{r}{0}\epsilon^0 + \binom{r}{1}\epsilon^1 + \binom{r}{2}\epsilon^2 + \cdots.$$

If ϵ is close to zero and r is less than one, then the third and each subsequent term are also close to zero. For example, if $\epsilon = 0.01$ and $r = 1/2 = 0.5$, then the third term is

$$\binom{r}{2}\epsilon^2 = \binom{0.5}{2}(0.01)^2 = \frac{0.5(0.5-1)}{2}(0.0001) = -0.000\,0125,$$

which is, indeed, very close to zero. The reader can verify that the fourth term is even closer to zero. Therefore, (3) can be written as

$$(1+\epsilon)^r = \binom{r}{0}\epsilon^0 + \binom{r}{1}\epsilon^1 + E = 1 + r\epsilon + E,$$

where

$$E = \sum_{n=2}^{\infty}\binom{r}{n}\epsilon^n.$$

If ϵ is close to zero, then

$$(4)\qquad (1+\epsilon)^r \approx 1 + r\epsilon,$$

since the E term is also close to zero. The E term can be regarded as the error in the approximation. In (4), let $r = 1/n$ and $\epsilon = x - 1$. By requiring that x be close to one, ϵ will necessarily be close to zero, and we obtain

$$(5)\qquad x^{1/n} \approx 1 + \frac{1}{n}(x-1).$$

High school students could be convinced of the validity of (5) by the following argument, which does not refer to the binomial series. If x is close to one, then the nth root of x is also very close to one. Similarly, for x close to one, the term

$$\frac{1}{n}(x-1)$$

is close to zero, and consequently,

$$1 + \frac{1}{n}(x-1)$$

is close to one. The result in (5) follows because both expressions $x^{1/n}$ and

$$1 + \frac{1}{n}(x-1)$$

are necessarily close to one when x is close to one.

Now let $x = a^{1/2^k}$. For large k, $x = a^{1/2^k}$ is close to one. For example, if $k = 15$, then x is the 32 768th root of a. Such a large nth root of a is very close to one regardless of the size of a (up to the capacity of most four-function calculators). Your students can check this by entering a large number in their calculators, say $a = 99\,999\,999$, and computing $a^{1/2^{15}}$ by depressing the square root key fifteen times. Then, by substituting $x = a^{1/2^k}$ in (5), we obtain

$$(6)\qquad (a^{1/2^k})^{1/n} \approx \left(\frac{a^{1/2^k}-1}{n} + 1\right).$$

Since

$$[(a^{1/2^k})^{1/n}]^{2^k} = a^{1/n},$$

raising both sides of (6) to the 2^k power yields the result of our theorem. Because $a^{m/n} = (a^{1/n})^m$, this method can also be used to compute rational roots of real numbers on a simple four-functional calculator with a square-root key and squaring capability.

The method used in this article can be applied by algebra teachers to stimulate a worthwhile discussion of approximation theory as well as to provide an application of the properties of exponents. Of course, the method also provides a useful way of computing rational roots if a calculator with an x^y key is not available.

REFERENCES

Eimer, Rebecca A. "Cube Roots on a Calculator." *Mathematics Teacher* 70 (February 1977): 175.

Apostal, Tom M. *Calculus*. New York: Blaisdell Publishing Co., 1962.

activities

Edited by **Evan M. Maletsky,** *Montclair State College, Upper Montclair, New Jersey*
Christian Hirsch, *Western Michigan University, Kalamazoo, Michigan*
Daniel Yates, *Mathematics and Science Center, Glen Allen, Virginia*

CALCULATING ORDER

By Susan Smith, Bel Air High School, El Paso, TX 79925

Teacher's Guide

Grade level: 7–10

Materials: Copies of student worksheets and calculators

Objective: To provide a calculator experience developing and reinforcing the concept of order of operations

Background: The conventional hierarchy for evaluating arithmetical expressions is (1) exponentiation, (2) multiplication and division, and (3) addition and subtraction. When parentheses are used, operations within them must be completed according to the hierarchy above before continuing. Calculators with algebraic entry, however, will complete operations as they are keyed.

Directions: Distribute worksheets and calculators.

Sheet 1: Explain the instructions and examples. Some forms are easier to calculate and can be entered in the calculator in the order written. Others will require intermediate answers or rearranging the terms before the correct answer can be obtained.

Sheet 2: Explain the instructions and example. Keying in the numbers and operations as they appear from left to right results in an answer (b) to a new expression (c). In the example, the keying order for (b) and (c) is

$$5 + 7 - 9 \times 3 + 6 \div 2 =$$

Sheet 3: Explain the instructions and example. The rewriting of the expression does not change the mathematical meaning but does allow the problem to be entered in the calculator in order from left to right.

Answers to all questions are given on p. 181.

CALCULATING ORDER SHEET 1

Instructions:

Calculate the answers for each of the following pairs of expressions using the conventional rules regarding the order of operations. Use calculator memory or write intermediate answers as needed.

Example:

(a) $7 + 4 \times 2 - 6 =$ 9

(b) $(7 + 4)2 - 6 =$ 16

1. (a) $8 \times 6 + 3 \times 5 =$ ______

   (b) $(8 \times 6 + 3)5 =$ ______

2. (a) $(12 + 4)5 + 8 =$ ______

   (b) $12 + 4 \times 5 + 8 =$ ______

3. (a) $72/16 - 4 + 31 \times 7 =$ ______

   (b) $(31 \times 7 + 72)/16 - 4 =$ ______

4. (a) $(13 \times 8 + 7)2 + 11 =$ ______

   (b) $11 + 2(13 \times 8 + 7) =$ ______

5. (a) $14 + 3(6 + 24)/2 =$ ______

   (b) $(6 + 24)3/2 + 14 =$ ______

6. (a) $(5 + (18 + 31)/16)24 =$ ______

   (b) $((18 + 31)/16 + 5)24 =$ ______

7. (a) $5(6 + 14(17 + 13)/4)/3 =$ ______

   (b) $((17 + 13)/4 \times 14 + 6)5/3 =$ ______

8. (a) $11(3 + 8(4 + 7)/3) =$ ______

   (b) $(4 + 7)8 + 3/3 \times 11 =$ ______

9. (a) $(((37 + 4)16 + 28)/3 - 36)/14 =$ ______

   (b) $37 + 4 \times 16 + 28/3 - 36/14 =$ ______

Instructions: SHEET 2

(a) Calculate answers according to the conventional rules regarding the order of operations. Use calculator memory or write intermediate answers as needed.

(b) Enter numbers and complete each operation in the order listed to calculate a second answer.

(c) Rewrite the expression, using parentheses as needed to indicate correct algebraic form for the answer you have calculated in (b).

Example:

5 + 7 - 9 x 3 + 6/2 = <u>-12</u> (a)

(b) <u>7.5</u> = <u>((5 + 7 - 9)3 + 6)/2</u> (c)

1. 2.2 x 3.71 + 4.035 x 5.6 = ______ (a)

(b) ______ = ____________________ (c)

2. 2.16 + 3.02 x 4.7 + 5.89 = ______ (a)

(b) ______ = ____________________ (c)

3. 5.7/3.8 - 6.4/2.008 = ______ (a)

(b) ______ = ____________________ (c)

4. 7.2 x 8.4 + 5.6/4 + 16.7 = ______ (a)

(b) ______ = ____________________ (c)

5. 4.9/5.1 x 6.3 + 4.8 x 7 = ______ (a)

(b) ______ = ____________________ (c)

6. 3.28 x 7.1 + 4.26 x 2.5 + 6.03/3.7/5.2 = ______ (a)

(b) ______ = ____________________ (c)

7. 4.26 x 5.9 + 9.8/7.6 - 3.4 = ______ (a)

(b) ______ = ____________________ (c)

8. 5.08 + 7.62/6.8 - 4.6/13.7 = ______ (a)

(b) ______ = ____________________ (c)

Instructions: SHEET 3

(a) Write the equivalent expression for the one given that will allow you to enter the numbers and operations in the order listed.

(b) Calculate the answer.

Example:

16 + 7(8 + 5)/4

(a) (8 + 5) 7/4 + 16 = 38.75 (b)

1. 4.7 + 21.4(7.8 + 6.7)

(a) ____________ = ____________ (b)

2. (19.8 + 32.6(25.9) - 7.5)/2

(a) ____________ = ____________ (b)

3. (5.4 + (9.02 + 2.48)/3.4) x 7.3

(a) ____________ = ____________ (b)

4. $6.42(7.03 + 4.66)^2$

(a) ____________ = ____________ (b)

5. 2.3(6.9(3.7 + 4.3/7.4) + 8.6)

(a) ____________ = ____________ (b)

6. (16.2 + 4(3.1)(4.2))/7.8

(a) ____________ = ____________ (b)

7. 3.6(7.9 + 5.2(4.2 + 9.2)/6)/5.3

(a) ____________ = ____________ (b)

8. ((17.4 + 4.2(3.5 + 5.3))6.7 - 7.6)/3 - 8.8

(a) ____________ = ____________ (b)

9. ((5.1 + 3.02(7.6 + 8.1)/8.62)/3.4)/4.2

(a) ____________ = ____________ (b)

activities

Edited by **Evan M. Maletsky,** *Montclair State College, Upper Montclair, New Jersey*
Christian Hirsch, *Western Michigan University, Kalamazoo, Michigan*
Daniel Yates, *Mathematics and Science Center, Glen Allen, Virginia*

CALCULATOR CAPERS

By Rosemary Schmalz, S.P., Saint Mary-of-the-Woods College,
Saint Mary-of-the-Woods, IN 47876

Teacher's Guide

Grade level: 7–11

Materials: Calculators and one set of worksheets for each student.

Objective: To provide an opportunity for students to discover numerical patterns.

Procedure: Sheets 1, 2, and 3 are independent of each other and can therefore be used on separate days. If there are only a limited number of hand-held calculators available you may wish to have students work on this activity in small groups. Provide each student with a copy of one or more of the worksheets. Be sure pupils understand the directions before they begin.

Supplementary activities: Students can be challenged to demonstrate why these patterns occur. They might also be encouraged to find other numerical patterns. To get them started you might have them investigate the pattern formed by the following products: 4×9, 44×99, 444×999, and so on. Other patterns can then be generated by replacing the 4 with a different single-digit number. A variety of additional patterns may be found in Carman and Carman (1970).

REFERENCE

Carman, R., and M. Carman. "Number Patterns." *Arithmetic Teacher* 17 (Dec. 1970): 637–39.

Calculator Capers

For each part, do the problems in the first column with a calculator. Then, by looking for patterns in the results, predict the answers to the problems in the second column without actually calculating them.

A.

(1 x 9) + 2 = __________	(12,345 x 9) + 6 = __________
(12 x 9) + 3 = __________	(1,234,567 x 9) + 8 = __________
(123 x 9) + 4 = __________	(________ x 9) + __ = 111,111,111

B.

(9 x 9) + 7 = __________	(9,876 x 9) + 4 = __________
(98 x 9) + 6 = __________	(9,876,543 x 9) + 1 = __________
(987 x 9) + 5 = __________	(________ x 9) + __ = 888,888

C.

(1 x 9) - 1 = __________	(4,321 x 9) - 1 = __________
(21 x 9) - 1 = __________	(7,654,321 x 9) - 1 = __________
(321 x 9) - 1 = __________	(________ x 9) - 1 = 5,888,888

D.

(1 x 8) + 1 = __________	(1,234 x 8) + 4 = __________
(12 x 8) + 2 = __________	(12,345,678 x 8) + 8 = __________
(123 x 8) + 3 = __________	(________ x 8) + __ = 987,654

E.

(99 x 1) + 1 = __________	(99 x 5) + 5 = __________
(99 x 2) + 2 = __________	(99 x 8) + 8 = __________
(99 x 3) + 3 = __________	(99 x __) + __ = 700

Calculator Capers

For each part, do the problems in the first column with a calculator. Then, by looking for patterns in the results, predict the answers to the problems in the second column without actually calculating them.

F.

99 x 9 = ________	99 x 5 = ________
99 x 8 = ________	99 x 2 = ________
99 x 7 = ________	99 x __ = 396

G.

37 x 3 = ________	37 x 15 = ________
37 x 6 = ________	37 x 24 = ________
37 x 9 = ________	37 x __ = 666

H.

999,999 x 2 = ________	999,999 x 6 = ________
999,999 x 3 = ________	999,999 x 9 = ________
999,999 x 8 = ________	999,999 x __ = 6,999,993

I.

15,873 x 7 = ________	15,873 x 35 = ________
15,873 x 14 = ________	15,873 x 56 = ________
15,873 x 21 = ________	15,873 x __ = 777,777

J.

3,367 x 3 = ________	3,367 x 15 = ________
3,367 x 6 = ________	3,367 x 21 = ________
3,367 x 9 = ________	3,367 x __ = 80,808

Calculator Capers

For each part, do the problems in the first column with a calculator. Then, by looking for patterns in the results, predict the answers to the problems in the second column without actually calculating them.

K.

$(11)^2$ = ____________	$(11{,}111)^2$ = ____________
$(111)^2$ = ____________	$(11{,}111{,}111)^2$ = ____________
$(1{,}111)^2$ = ____________	$(\_\_\_\_\_\_\_\_\_\_\_\_)^2$ = 12,345,654,321

L.

$(34)^2$ = ____________	$(33{,}334)^2$ = ____________
$(334)^2$ = ____________	$(33{,}333{,}334)^2$ = ____________
$(3{,}334)^2$ = ____________	$(\_\_\_\_\_\_\_\_\_\_\_\_)^2$ = 111,111,555,556

M.

$(67)^2$ = ____________	$(66{,}667)^2$ = ____________
$(667)^2$ = ____________	$(6{,}666{,}667)^2$ = ____________
$(6{,}667)^2$ = ____________	$(\_\_\_\_\_\_\_\_\_\_\_\_)^2$ = 4,444,444,488,888,889

N.

$(98)^2$ = ____________	$(99{,}998)^2$ = ____________
$(998)^2$ = ____________	$(99{,}999{,}998)^2$ = ____________
$(9{,}998)^2$ = ____________	$(\_\_\_\_\_\_\_\_\_\_\_\_)^2$ = 999,996,000,004

O.

$(49)^2$ = ____________	$(49{,}999)^2$ = ____________
$(499)^2$ = ____________	$(4{,}999{,}999)^2$ = ____________
$(4{,}999)^2$ = ____________	$(\_\_\_\_\_\_\_\_\_\_\_\_)^2$ = 24,999,990,000,001

Edited by **Evan M. Maletsky,** *Montclair State College, Upper Montclair, New Jersey*
Christian Hirsch, *Western Michigan University, Kalamazoo, Michigan*
Daniel Yates, *Mathematics and Science Center, Glen Allen, Virginia*

CALCULATOR CROSSWORD PUZZLE

By Joseph F. Goodhue, Newport High School, Newport, NH 03773

Teacher's Guide

Grade Level: 7–12.

Materials: Eight-place pocket calculator (preferably with a memory).

Objectives: In this activity students use pocket calculators to demonstrate their knowledge of the correct order of operation for a given mathematical expression.

Directions: The first worksheet provides some warmup exercises that review the rules for the order of operations and establishes the procedure for spelling out words by inverting the calculator. In general, multiplication and division take priority over addition and subtraction unless symbols of grouping (parentheses, brackets, and braces) are involved. The rule for the order of operations is as follows:

First carry out all operations within symbols of grouping. Then perform multiplications and divisions in the order in which they occur from left to right. Finally, perform all additions and subtractions.

For each entry in the crossword puzzle (sheet 2), the student performs the indicated computation, gets a result, and then inverts the calculator to read a word or abbreviation. The word or abbreviation is checked against the verbal cue and then entered into the appropriate cells in the crossword puzzle diagram.

Looking at 1 Across on the crossword puzzle (sheet 2), we need a four-letter word for "legendary Viking." Entering the expression $5(200 \times 15 - 10 \times 172 - 6)$ on the calculator in the correct order of operation should produce the number

6 3 7 0.

Inverting the calculator, we read the name

0 L E 9,

which we write in the given spaces in our puzzle.

The answer sheet (sheet 3) can be reproduced for the students if desired.

REFERENCE

Vannatta, Glen D., A. Wilson Goodwin, and Harold P. Fawcett. *Algebra Two: A Modern Course.* Columbus, Ohio: Charles E. Merrill, 1966.

ORDER OF OPERATIONS

Sheet 1

1. In each of the following, use a calculator and the rule for the order of operations to compute the given expression. When you get an answer, turn your calculator upside down and read the word. Record intermediate steps and the word.

Example: 123(45 + 72) - 40584 ÷ 6 + 13 x 7

= 123 x 117 - 40584 ÷ 6 + 13 x 7

= 14391 - 6764 + 91

= 7718

B I L L

a) (10+13(236+587)-595x5)÷10 000

b) 37+(20+(19+13x41-5))

c) 96(78-24÷8-25)+24x25-91

d) 623x411-213x303+1296÷4x579-288-16

2. Write the letters that correspond to the numerals:

0	1	2	3	4	5	6	7	8	9
_	_	_	_	_	_	_	_	_	_

3. Make your calculator read these words:

	Keys to Press
SIZE	____________
gIggLE	____________
hOBBLE	____________
BOBSLED	____________

CALCULATOR CROSSWORD PUZZLE Sheet 2

ACROSS

1. Legendary Viking
5(200x15-10x172-6)

4. East African river
5(115889x3+2)

9. Actor Hope
(12x199+36)÷3

11. African desert
(350x20+40x5)÷4+6

12. Rumanian river
15(4x895+3x3)

13. Siberian river
(14x20x5+40)÷(5x3+3)

15. Actor Greene (initials)
2x15-5x2+5x8+7

18. _____ Horizonte
0.8856÷12

20. Swiss lake
2(9x7x58+5)

22. Filmmaker Goldwyn (initials)
(38x14+11x4)÷8-8+1

24. Actor Bridges
(16.6-6.2)÷(5x30-20)

27. Belgian river
25x20(35x2+1)+37

28. Dutch painter
8x25(4x5+3)+6

31. State
(50(7x4+5))÷3+221

32. Nigerian Mountain Range
15(300x320+1223)

33. Make dirty
5x27(9x7-10)-50

DOWN

1. French river
10((70x10)÷2+1)

2. _____ of Reims
4x0.0220+0.0003

3. Nazi minister (WW II)
88888888-31500888+150x2+6

5. German sculptor
(313254+305505+5x29)+8

6. Actor Bridges
0.80÷16+0.03

7. Hitler's mistress (initials)
8x7+9x3

8. Belgian University
36007+15.5x20

10. College Degree
(350+32x0.3)÷6.2

14. Self 0.08x0.9-0.009

16. Same as 3 Down

17. Old Saxon (abbr.)
(9x39-1)÷7

19. Old Latin (abbr.)
3(8x21-4)÷6+1-13

20. Little ____ Horn
22x32-16x6+10

21. Capt. of H.M.S. *Bounty*
2((14x12+3)x(15x9)+4

23. Belonging to 28 Across
6(2733÷3x10-9)÷10000

25. Norwegian city
0.1230-0.0340-0.0140

26. Telephone inventor
15+8x8x11x11-21

27. Long Island (abbr.)
(21x14-5)÷17

29. Old English (abbr.)
(645÷15+47)÷3

30. Mathematician Boole (initials)
(12x79-2)÷(30÷6+6)

ANSWER SHEET

Sheet 1:

1. a) 0.7734 BILL   b) 604 hOg
   c) 5309 GOES   d) 378806 gOBBLE

2.

0	1	2	3	4	5	6	7	8	9
O	I	Z	E	h	S	g	L	B	G
or D									

3. SIZE 3215
   gIggLE 376616
   hOBBLE 378804
   BOBSLED .375808

Sheet 2:

activities

Edited by **Evan M. Maletsky,** *Montclair State College, Upper Montclair, New Jersey*
Christian Hirsch, *Western Michigan University, Kalamazoo, Michigan*
Daniel Yates, *Mathematics and Science Center, Glen Allen, Virginia*

CALCULATOR GRAPHING

By William A. Miller and Donald W. Hazekamp
Central Michigan University, Mount Pleasant, MI 48859

Teacher's Guide

Grade Level: 7–12

Materials: One set of activity sheets and a hand-held calculator for each student. If you do not have an adequate supply of calculators, students may work in pairs.

Objectives: To provide experience in using calculators to compute squares, reciprocals, and square roots. Experience also is gained in rounding numbers to the nearest hundredth, in plotting points, and in drawing graphs.

Directions: Distribute the activity sheets, one at a time, to each student. Have the students complete sheet 1 before beginning sheet 2.

Sheet 1. This initial activity familiarizes students with using a calculator to compute squares of numbers from zero to one and in rounding off these numbers to the nearest hundredth. A nice extension is to graph the curve $y = x^3$ in the same interval. Some students may be bothered by the fact that the cubes of 0.00, 0.05, 0.10, and 0.15 are each 0.00 when rounded off to the nearest hundredth.

Sheet 2. This activity provides more practice in using the calculator, in rounding off, and in graphing. If your calculators do not have the reciprocal function, these values can be computed by using the division function ($y = 1 \div x$).

Sheet 3. This activity requires calculators with a square-root key. It may be helpful to have your students graph the curve $y = \sqrt{x}$ in the interval $0 \le x \le 1$ before working on this sheet. In completing this activity, have your students round off x^2 to the nearest hundredth. The corresponding values of $\sqrt{1 - x^2}$ are sufficiently accurate when computed from that data. With students who have the appropriate background, you may wish to develop the relationship between the equations $x^2 + y^2 = 1$ and $y = \sqrt{1 - x^2}$. For further study, students may be interested in comparing the graphs of $y = \sqrt{4 - x^2}$ or $y = \sqrt{9 - x^2}$ on the corresponding intervals $0 \le x \le 2$ or $0 \le x \le 3$. For these graphs, change the horizontal scale so that ten to twenty points are generated.

BIBLIOGRAPHY

Miller, William A. *Laboratory Activities in Algebra.* Teacher ed. Portland, Maine: J. Weston Walch, Publisher, 1974, pp. T28–T30 and 52–57.

SHEET 1

Graphing $y = x^2$

1. Use your calculator to complete the following table. Round off your answers to the nearest hundredth.

x	$y = x^2$
.00	
.05	
.10	
.15	
.20	
.25	
.30	
.35	
.40	
.45	
.50	
.55	
.60	
.65	
.70	
.75	
.80	
.85	
.90	
.95	
1.00	

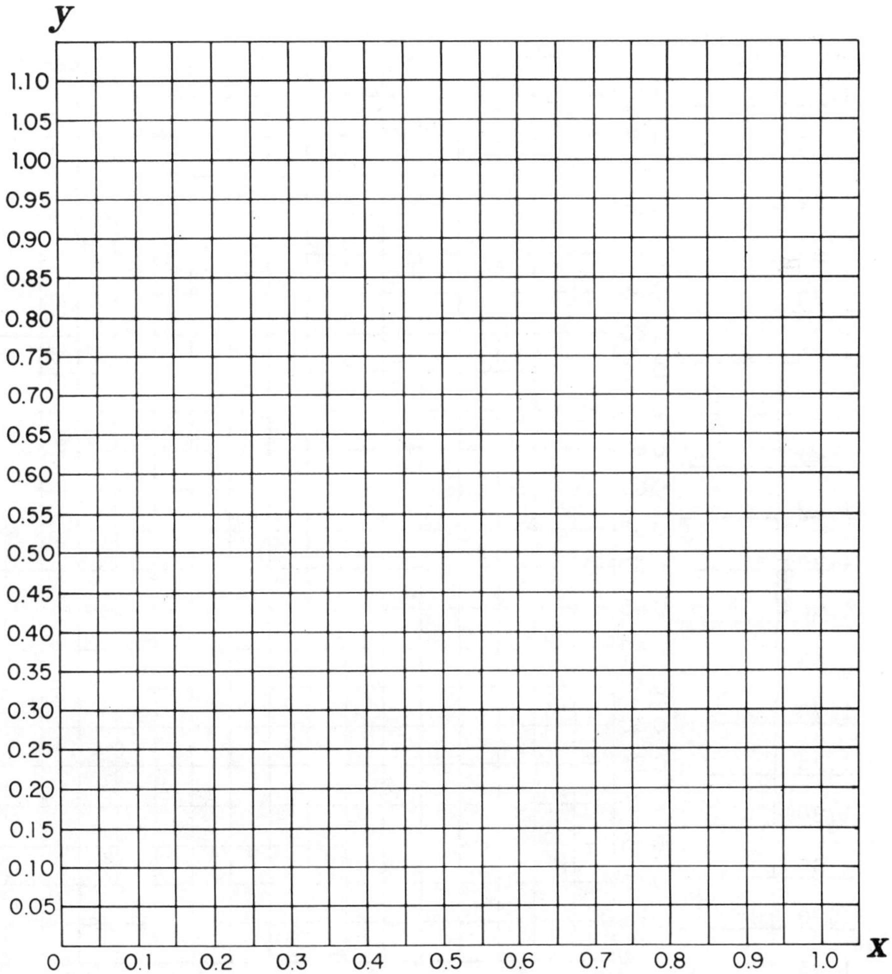

2. After you have completed the table, graph the ordered pairs (x , y).

3. Connect the points with a smooth curve to complete the graph.

Graphing $y = \frac{1}{x}$ from 1 to 2

1. Use your calculator to complete the following table. Round off your answers to the nearest hundredth.

x	$y = \frac{1}{x}$
1.00	
1.05	
1.10	
1.15	
1.20	
1.25	
1.30	
1.35	
1.40	
1.45	
1.50	
1.55	
1.60	
1.65	
1.70	
1.75	
1.80	
1.85	
1.90	
1.95	
2.00	

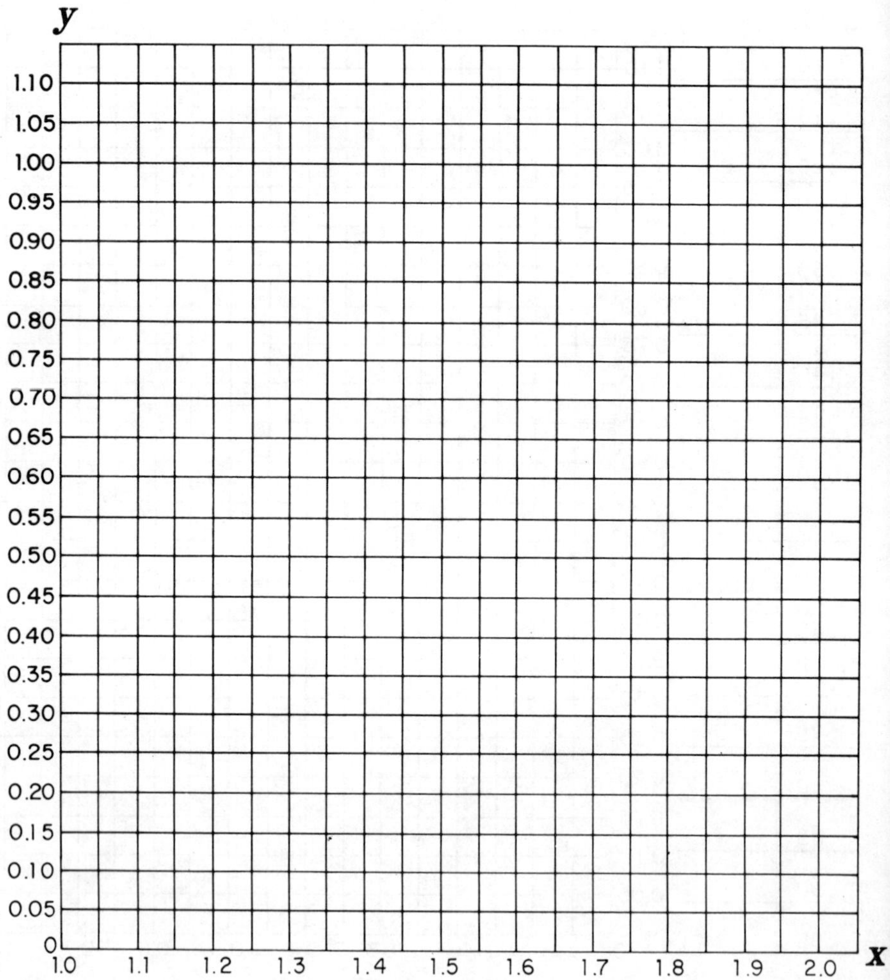

2. After you have completed the table, graph the ordered pairs (x , y).

3. Connect the points with a smooth curve to complete the graph.

Graphing $x^2 + y^2 = 1$

. Use your calculator to complete the following table. Round off your answers in each column to the nearest hundredth.

x	x^2	$1-x^2$	$y=\sqrt{1-x^2}$
.00			
.05			
.10			
.15			
.20			
.25			
.30			
.35			
.40			
.45			
.50			
.55			
.60			
.65			
.70			
.75			
.80			
.85			
.90			
.95			
.00			

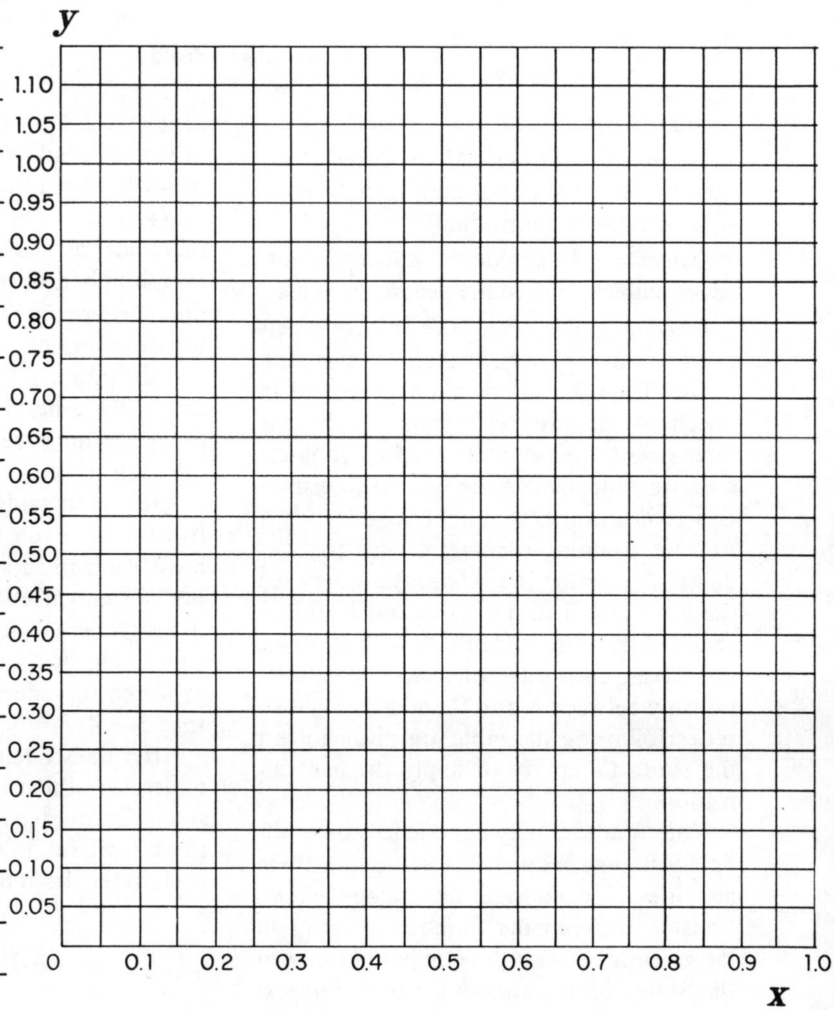

:. After you have completed the table, graph the ordered pairs (x , y).

•. Connect the points with a smooth curve to complete the graph.

activities

Edited by **Evan M. Maletsky,** *Montclair State College, Upper Montclair, New Jersey*
Christian Hirsch, *Western Michigan University, Kalamazoo, Michigan*
Daniel Yates, *Mathematics and Science Center, Glen Allen, Virginia*

PATTERN GAZING

By Cherie Adler Aviv, University of Georgia, Athens, GA 30602

Teacher's Guide

Grade Levels: 7–12

Materials: One set of worksheets for each student, a transparency of each worksheet, calculators (optional).

Objectives: To enable the students to discover patterns and make generalizations.

Directions: Distribute the activity sheets one at a time. With each sheet, discuss with the students the information presented in the charts. Be sure they understand that the three dots (. . .) represent unspecified intermediate addends. You may need to discuss how to determine the number of addends N without counting each term. Ask the students to examine the relationship between the last number in the series and N of that series. After the students have completed the tables, encourage them to seek a relationship between N and the sum S. The two series following the table are given to test the student's ability to apply the generalization.

Your students may experience some difficulty in determining N for the first three activities. If so, consider discussing the relationship between the number of terms in the given series with the number of terms in the series of counting numbers. For example, compare the number of terms in the series $1 + 3 + 5$ with $1 + 2 + 3 + 4 + 5 + 6$. Students should recognize that the second series has six terms and can be regrouped as $(1 + 3 + 5) + (2 + 4 + 6)$. Since there are as many odd-number terms as even-number terms, the number of odd-number terms will always be half the number of the terms in the second series. Additional examples of this type may be necessary. This generalization will prove helpful in determining the sum of the series following each table.

After each table, the students are asked to write their rule in words or symbols. For most students, a statement such as "multiply N times itself" for (1) or "multiply N times $(N + 1)$" for (3) should suffice. Advanced algebra students, however, might be expected to generalize in (3), for example, that $2 + 4 + 6 + \cdots + (2n) = n(n + 1)$.

After the students have completed all the tables, ask them to look back at their results. Encourage the students to see how their rules for the various tables relate to each other. In activity 4, the rule is

$$S = \frac{N(N + 1)}{2}.$$

The rule for activity 6 is

[*Continued on p. 162*]

PATTERN GAZING

(1) Complete this table to see what happens when we add odd numbers.

Series	No. of Addends (N)	Sum (S)
1	1	1
1 + 3	2	4
1 + 3 + 5		
1 + 3 + 5 + 7		
1 + 3 + 5 + 7 + 9		
1 + 3 + 5 + 7 + 9 + 11		

Can you find a way to calculate the sum without adding each term? If so, use your rule to calculate these sums:

$1 + 3 + \cdots + 17 + 19 =$ ____________________

$1 + 3 + \cdots + 23 + 25 =$ ____________________

Write your rule in words or symbols. ____________________

____

(2) Let's see what happens when we alternately add and subtract odd numbers. Complete the table.

Series	No. of Addends (N)	Sum (S)
1	1	1
1 - 3	2	-2
1 - 3 + 5		
1 - 3 + 5 - 7		
1 - 3 + 5 - 7 + 9		

Can you find a way to calculate the sum without alternately adding and subtracting each term? If so, use your rule to calculate these sums:

$1 - 3 + \cdots + 17 - 19 =$ ____________________

$1 - 3 + \cdots + 41 - 43 =$ ____________________

Write your rule in words or symbols. ____________________

____

____

PATTERN GAZING

(3) Let's try adding even numbers.

Series	No. of Addends (N)	Sum (S)
2	1	2
2 + 4	2	6
2 + 4 + 6		
2 + 4 + 6 + 8		
2 + 4 + 6 + 8 + 10		
2 + 4 + 6 + 8 + 10 + 12		

Can you find a way to calculate the sum without adding each term? If so, use your rule to calculate these sums:

2 + 4 + 6 + ··· + 26 + 28 = ____________

2 + 4 + 6 + ··· + 48 + 50 = ____________

Write your rule in words or symbols.____________

____

(4). Let's see what happens when we add both odd and even numbers.

Series	No. of Addends (N)	Sum (S)
1	1	1
1 + 2	2	3
1 + 2 + 3		
1 + 2 + 3 + 4		
1 + 2 + 3 + 4 + 5		

Can you find a way to calculate the sum without adding each term? (Hint: Compare chart 4 with chart 3)

If so, use your rule to calculate these sums:

1 + 2 + ··· + 17 + 18 = ____________

1 + 2 + ··· + 43 + 44 = ____________

Write your rule in words or symbols.____________

____

____

PATTERN GAZING

(5) Let's see what happens when we square first and then alternately add and subtract.

Series	No. of Addends (N)	Sum (S)
1^2	1	1
$1^2 - 2^2$	2	-3
$1^2 - 2^2 + 3^2$		
$1^2 - 2^2 + 3^2 - 4^2$		
$1^2 - 2^2 + 3^2 - 4^2 + 5^2$		

Can you find a way to calculate the sum without alternately adding and subtracting each term? (HINT: Compare chart 5 with chart 4)

If so, use your rule to calculate these sums:

$1^2 - 2^2 + \cdots + 17^2 - 18^2 =$ ______________________

$1^2 - 2^2 + \cdots - 24^2 + 25^2 =$ ______________________

Write your rule in words or symbols.______________________

____

(6) What about adding the cubes of the counting numbers? Complete the table.

Series	No. of Addends (N)	Sum (S)
1^3	1	1
$1^3 + 2^3$	2	9
$1^3 + 2^3 + 3^3$		
$1^3 + 2^3 + 3^3 + 4^3$		
$1^3 + 2^3 + 3^3 + 4^3 + 5^3$		

Can you find a way to calculate the sum without adding each term? (HINT: Compare chart 6 with chart 4).

If so, use your rule to calculate these sums:

$1^3 + 2^3 + \cdots + 13^3 + 14^3$? ______________________

$1^3 + 2^3 + \cdots + 19^3 + 20^3$? ______________________

Write your rule in words or symbols.______________________

____

$$S = \left[\frac{N(N+1)}{2}\right]^2.$$

This latter result says that the sum of the first N cubes is the square of the sum of the first N natural numbers.

Most students should be able to substantiate the relationship between charts 3 and 4. In chart 3, $2 + 4 + 6 + \cdots + (2n) = n(n+1)$, and in chart 4, $1 + 2 + 3 + \cdots + n = n(n+1)/2$. A statement such as "each term in the series in chart 4 is half the corresponding term in the parallel series in chart 3" should suffice. Students familiar with factoring may see the relationship as $2 + 4 + 6 + \cdots + (2n) = 2(1 + 2 + 3 + \cdots + n)$.

Follow-up Activities: Examine these series for similar patterns:

(1) $2 - 4 + 6 - 8 + \cdots$
(2) $1 - 2 + 3 - 4 + \cdots$
(3) $1^2 + 2^2 + 3^2 + \cdots$

Answers:

1. $10^2 = 100$; $13^2 = 169$; $s = n^2$
2. -10; -22; $S = \begin{cases} N \text{ if } N \text{ is odd} \\ -N \text{ if } N \text{ is even} \end{cases}$
3. $(14)(15) = 210$; $(25)(26) = 650$; $S = N(N+1)$
4. $\frac{(18)(19)}{2} = 171$; $\frac{(44)(45)}{2} = 990$; $S = \frac{N(N+1)}{2}$
5. -171; 325; $S = \begin{cases} N(N+1)/2 \text{ if } N \text{ is odd} \\ -N(N+1)/2 \text{ if } N \text{ is even} \end{cases}$
6. $\left[\frac{(14)(15)}{2}\right]^2 = 11025$; $\left[\frac{(20)(21)}{2}\right]^2 = 44100$; $S = \left[\frac{N(N+1)}{2}\right]^2$

CALCULATOR CHARADES

Calculators can be fun as well as practical.

By PATRICK J. BOYLE
Santa Rosa Junior College
Santa Rosa, California 95401

IN THE summer of 1974, *Time* magazine had a short article about using the new pocket-sized electronic calculator to form words when the display is inverted. Since that time, I've seen several other mentions of this activity—an excellent article appeared in the April 1975 issue of *Datamation*. Another such example appeared in our local newspaper, the *Santa Rosa Press-Democrat,* by columnist Gaye LeBaron. It is probably much like ones you may have seen.

Calculated to keep you in . . . Suspense. If you have access to a pocket calculator, humor me and play this silly game.

Punch 142 for the number of Israeli soldiers. Now punch 154 for the number of Arabs. Don't add or subtract or anything, just make a continuous number. Now punch 69 for the length of the battlefront. Now hit the times (×) sign and a five for the Five Day War. Push the equal (=) button. Turn the computer upside down and see who won.

Such activity is recreational and gives practice using the devise. Moreover, it helps to demystify the strange, powerful, little packages for many students who feel threatened by a calculator. Consider the following problem, and you may find an opportunity to involve students in a similar activity.

1. Use a calculator to describe the number $\{(16\cdot 599)^2 - (29\cdot 59)\}$.

Note that the calculation requires the use of a memory on most machines or the student must write down an intermediate result and reenter it. In any event, if you calculated it, you found it to be 91851345; and inverted, it displays SHEISBIG, which uses up the eight-digit display of most machines. For larger displays, one could use 9185.345 or 91834551.0 and devise appropriate calcuations to obtain them.

A story is an interesting way to embellish the calculation; it makes a game out if it, sort of a charade using calculators.

2. Traveler Ted was complaining that flight 239 no longer leaves at 9 A.M. He could take flight 261, which leaves two hours later and costs nine dollars more. Ted decided to use his pocket calculator to help him decide which flight to take. Follow his steps and discover the advice of the machine.
   a. Take the second flight number and square it.
   b. Add the first flight number.
   c. Multiply by the rate increase.
   d. Subtract the second departure time.
   e. Multiply by the first flight time.
   f. Turn the calculator upside down.

A larger display would permit 55370677.1, a nicer reply. But notice that you would need a new set of calculations to lead to this answer. Note also that information must be taken from the problem, much the same as in any word problem. It would be easy enough to make this process more involved. That is, to hide some of the data and provide some extraneous data. It could then provide practice in excerpting the proper data from a problem statement.

The calculator need not be inverted to display a message once you have learned to recognize the alphabetic characters. Who's in charge, after all?

3. Do you rule the calculator, or does the little black box rule you? Carry out the following and find out!
   a. Square eighty.
   b. Add the square of seven.
   c. Multiply by three-squared.
   d. Multiply by five.
   e. Multiply by seventeen.
   f. Multiply by one hundred sixty-three.

Don't invert, the calculator should be announcing its resignation. The answer is somewhat sexist, but after all, there are a limited number of symbols. All other examples require that the display be inverted.

This last example illustrates a good point. I discovered that 804 158 055 has factors of 3, 5, 17, 163, and 6449. The proper use of the exercise could lead to independent experimentation by the student. Finding the message is simple, given the directions. The calculator takes care of that, when used correctly. Disguising it in a story or a complicated series of calculations requires the message writer to examine the numbers very carefully. It is similar to a word problem, in that if you are given the equation, finding the answer is simple. Make the student the message writer. Limit the directions to certain operations, or require that a certain operation or function be included. After all, 158048055, or is the calculator the 8198055?

Here are a number of questions and "answers" that may give you some ideas for your particular classes. Watch out for round off and the differences in algorithms between various brands and models.

4. How is Math this year?
   $2^4 \cdot 3 \cdot 5 \cdot 83 \cdot 277$
5. How do you plan to get an A?
   $\{(197)(25.08) + 0.14\}(19)$
6. How was your shopping trip?
   $2 \cdot 17^2\{1261^2 - (9)(10)(11)\}$
7. What kind of work does Kinney do?
   $(5)(23)(53^2)(16421)$
8. What did you buy from him?
   $((((11.4)^2 + 0.16)^2 + 0.1856)^2 + 1)(2)$

And the last three are for those of you with a complete set of functions.

9. I need a volunteer!
   $10^{-1.01} - 1.3722 \cdot 10^{-5}$
10. Where is he in Alaska?
   $88^2 - e^{3.4} - 10^{-1.553}$
11. How is your math teacher's board work?
   $\sin(22.222°) - 2.4966 \cdot 10^{-6}$

Try this light touch of *Calculator Charades* and see if it doesn't spark some interest. But most importantly, watch closely for unexpected results!

REFERENCES

"Games Calculators Play." *Time,* 24 June 1974, p. 56.

Haberman, Jules, and Sergio Berstein. "Top Down Calculating." *Datamation,* April 1975, pp. 97–99.

LeBaron, Gaye. *Santa Rosa Press-Democrat,* 17 January 1975.

Answers

1.	91851345	SHEISBIG
2.	5537061	ILLGOLESS
3.	804158055	BOYISBOSS
4.	5517840	OHBLISS
5.	93877.1	I'LLBEG
6.	918517718	BILLISBIG
7.	5304557735	SELLSSHOES
8.	573344614	HIGHHEELS
9.	.09771	ILLGO'
10.	7714.00791	IGLOO'HILL
11.	.378193771	ILLEGIBLE'

The calculator a calendar?

HENRY F. KAHN

In addition to being head of the mathematics department at Thomas A. Edison High School in Philadelphia, Henry Kahn has been active in mathematics curriculum development. He has taught mathematics from elementary through undergraduate school.

Without a doubt, there are free calendars available from many quarters, and even the perpetual calendars can be obtained free or for modest sums. Thus, the electronic calculator is neither the cheapest nor necessarily the most efficient calendar. However, using it as a calendar will demonstrate some important mathematical concepts that are usually taught in the abstract.

Calendars are generally used to determine the day of the week of a certain date. If we wanted to know the day on which we celebrate Christmas 1976, we simply look at one of those little cards that many of us carry in our wallets. But if we wanted to know the day of the week on which December 25, 1989 falls, we would either have to consult a perpetual calendar or decide the result may not be worth the effort. And were we asked on what day Christmas 1776 was celebrated, we might reply sheepishly that General Washington was rather busy getting ready to cross the Delaware that night and the citizenry of our newly formed republic were probably too preoccupied to celebrate that year anyway!

In our discussions we refer of course to the Gregorian calendar that was instituted under the direction of Pope Gregory XIII in 1582 to correct the difference between solar and calendar time, which had accrued on the hitherto used Julian calendar. In fact, ten days had to be "deleted" in October 1582 to bring the calendar into close proximity with solar time. Thus, the system we develop here to determine the day of the week for any date will obviously be invalid for any date prior to about 1600. Furthermore, universal adoption of the Gregorian or "new" calendar took a long time throughout the Western World. England, for example—and with her, her colonies—did not adopt the new calendar until the mid seventeen hundreds. Hence, some discrepancies may arise in days of the distant past.

Since each week has seven days, the calendar is a practical application of the modulo 7 system. Briefly, this means that

$$n + r = 7k ,$$

where $n, k \in I$, the set of integers, and $r \in \{0, 1, 2, 3, 4, 5, 6\}$. Generally, the above equation is expressed

$$n \equiv r (\text{modulo } 7).$$

The foregoing statement, called a congruence, shows the remainder r when the integer n is divided by 7. This is fairly easy to determine when we use the division algorithm. But alas, "long division," even with single-digit divisors, is not a favorite activity of most of our students. And with the advent of the minicalculator, long division might go the way of the square root algorithm, or even the once popular cube root algorithm.

The calculator, however, does not show remainders. It only displays the quotient. But since the calculator continues the division process beyond the integer portion, we can determine the remainder from the tenth digit alone in the modulo 7 system:

$1/7 = .\overline{142857}$ $\quad 4/7 = .\overline{571428}$

$2/7 = .\overline{285714}$ $\quad 5/7 = .\overline{714285}$

$3/7 = .\overline{428571}$ $\quad 6/7 = .\overline{857142}$

Thus we note a 1 in the tenth place indicates that the remainder is 1, a 2 in the tenth place indicates the remainder is 2, a 4 indicates a remainder is 3, a 5 indicates the remainder is 4, a 7 indicates the remainder is 5, and 8 indicates that the remainder is 6.

We are now almost ready to use the calculator to determine the day of the week of any date. But we need one *reference* day. Almost any day may be used as such (except February 29), but choosing January 1 as the reference day will simplify this discussion. So we sneak a look at our pocket calendar and if it is a 1976 calendar we note that January 1 was a Thursday. We now set up a correspondence between r and the days of the week as follows:

1	2	3	4	5	6	0
↕	↕	↕	↕	↕	↕	↕
Thurs.	Fri.	Sat.	Sun.	Mon.	Tues.	Wed.

Suppose we wanted to know the day of July 4, 1976. We need to know the day of the *year* of July 4. The calculator helps us to quickly find that July 4 is day number 186 since

January has	31 days
February has	29 days (1976 is a leap year)
March has	31 days
April has	30 days
May has	31 days
June has	30 days
July has	4 days to the target day
	186

Hence, July 4 is day number 186 and $186/7 = 26.\overline{571428}$. But we are only interested in the tenth digit, which is a 5. This indicates that the remainder of the division problem is a 4, and 4 corresponds to Sunday. Thus, July 4, 1976 was a Sunday. This may not be a startling revelation, but suppose we wanted to know the day of the week on which we celebrate Christmas 1989? A 1976 calendar is of no help to us in this case; the calculator is, however.

Since a common year has 365 days, January 1 of the following year is day 366, and $366/7 = 52.285714$, which indicates that $r = 2$. Thus, each date of a succeeding year is one more than that of the previous year. Hence, we add one day for each year between 1989 and 1976. An additional day has to be added for each leap year in that interval. The procedure to determine the day of the week of December 25, 1989 is as follows:

1. December 25 in a common year is day number — 359
2. 1989 − 1976 — 13
3. Leap years are 1976, 1980, 1984, and 1988 — 4
4. Sum of the day of the year, the number of years, and the number of leap years — 376
5. $376/7 = 53.7 \ldots$, which indicates a remainder of 5.
6. Since 5 corresponds to Monday, December 25, 1989 is a Monday.

A similar process is used to determine the day of a date in a prior year. Suppose we wanted to know the day that Pearl Harbor was attacked on December 7, 1941. The problem is solved as the one above, except that the sum of the number of years and leap years is substracted from the number of the day in the year:

1. December 7 in a common year is day number — 341
2. 1976–1941 — 35
3. Leap years are 1944, 1948, 1952, 1956, 1960, 1964, 1968, 1972 — 8

   Sum: 43
4. Subtract sum of years and leap years — −43

   Result: 298
5. $298/7 = 42.5 \cdots$, which indicates a remainder of 4.
6. Since 4 corresponds to Sunday, December 7, 1941 was a Sunday.

Next, let us determine the day of the week of January 5, 1950. We note that the difference is a negative number:

1. January 5 is day number	5	
2. 1976–1950		26
3. Leap years are 1952, 1956, 1960, 1964, 1968, and 1972		6
		32
4. Subtract the sum of years and leap years	−32	
	−27	

5. (−27)/7 = −3.8 ⋯, which indicates a −6 remainder

This negative remainder requires the discussion to return briefly to modular arithmetic. We explain to students that in ordinary arithmetic −1 is the additive inverse of 1; 6 is the additive inverse of 1 in modulo 7, since 6 + 1 ≡ 0 (modulo 7). Since −1 in ordinary arithmetic and 6 in modulo 7 arithmetic are each the respective unique additive inverse of 1, it follows that −1 ≡ 6 (modulo 7). Similarly,

$$\begin{aligned} -2 &\equiv 5 \text{ (modulo 7)} \\ -3 &\equiv 4 \text{ (modulo 7)} \\ -4 &\equiv 3 \text{ (modulo 7)} \\ -5 &\equiv 2 \text{ (modulo 7)} \\ -6 &\equiv 1 \text{ (modulo 7).} \end{aligned}$$

Returning now to the problem of January 5, 1950,

6. −6 ≡ 1 (modulo 7); thus r = 1, which corresponds to Thursday.

Finally, let us discuss leap years. If the last two digits of a year date are divisible by 4, the year is a leap year. However, to adjust to a constantly changing discrepancy between calendar and solar time, century years are leap years only if their date is divisible by 400. Thus, while 1600 was a leap year, 1700, 1800, and 1900 were common years. The year 2000 will be a leap year.

With this information, we can now use the calculator to determine the day of the week of Christmas 1776:

1. December 25 is day number	359	
2. 1976–1776		200
3. Number of leap years in interval		47
		247
4. Subtract sum of years and leap years	−247	
	112	

5. 112/7 = 16.0, hence r = 0
6. Since 0 corresponds to Wednesday, Christmas 1776 was on a Wednesday.

This type of work can be very motivating, not only to the better students, but to those whose interests lie outside mathematics and who question the introduction of new concepts with the familiar "what good is this?"

EDITOR'S NOTE: For another article on calendars, see "Let's Play Mod 7," by Christopher E. Niemann in the May 1976 issue of the *Arithmetic Teacher*, pp. 348–50.

MYSTICISM AND MAGIC IN THE NUMBER SQUARES OF OLD CHINA

Numerology and the use of minicalculators can be combined to provide fascinating classroom activities.

By FRANK SWETZ

The Pennsylvania State University
Middletown, PA 17057

For centuries, the investigation of magic squares and their properties has intrigued both amateur and astute mathematicians. Considered today as mere recreations, these numbered squares can provide a boundless source of classroom exercises in numerical pattern recognition, mathematical exploration, and actual problem solving. The form of problem solving encountered in working with magic squares, such as completing a partially prescribed magic square or actually devising one according to a set of constraints, is particularly suited to, and indeed encouraged by, the use of a minicalculator. A knowledge of magic squares would be a helpful resource for teachers seeking enrichment exercises that can be associated with the use of a handheld calculator. Although much has been written on the general topic of magic squares, two fundamental aspects of this subject have received little attention in the popular literature: the origin of the magic square and the nature of its original "magical" properties. In this article we shall investigate these two facets of the history of the magic square and in doing so, supply useful background information, concepts, and procedures for stimulating students' interest in the subject.

Numerology and Mathematics: A Creative Union

Contained among the legends of old China is the story of how Emperor Yu of the Shang dynasty (2000 B.C.) received a tortoise-borne diagram—a diagram believed to contain the basic principles of mathematics. This incident was supposed to have taken place along the Lo River, a branch of the Yellow River; thus the diagram has come to be known as the *Lo shu* ("Lo River document"). When translated from its archaic form, the Lo shu is recognized as being a third-order magic square (fig. 1).

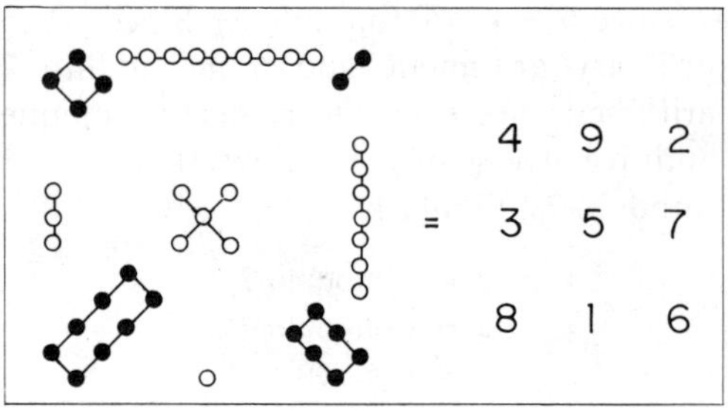

Fig. 1

Although its antiquity does not reach to the mythological Shang period, its existence can theoretically be traced to the Warring States era of Chinese history (403–221 B.C.), and the first specific written mention of it is found in the *Ta Tai li chi* [Records of rites compiled by Tai the Elder], a product of the first century B.C. The Lo shu is acknowledged by historians of mathematics to be the first known magic square. Thus, magic squares originated in China about the fourth century B.C.

From their inception in pre-Christian times until well into the tenth century, magic squares were indeed magical devices incorporated into amulets and charms and employed in divination rites. Vestiges of this influence can still be found in Tibetan astrological charts and diagrams. The religious and philosophical rationalizations used by the Chinese to interpret the inner workings of their squares were based on the dualistic metaphysical theory of the *Yin* and *Yang* devised during the Han dynasty. According to this cosmological system, the universe is ruled by Heaven through means of a process called the *Tao* ("the Universal Way"). Heaven, acting through the Tao, expresses itself in the interaction of two primal forces, the Yin and the Yang. The Yang, or male force, was a source of heat, light, and dynamic vitality and was associated with the sun; in contrast, the Yin, or female force, flourished in darkness, cold, and quiet inactivity and was associated with the moon (obviously the old Chinese were male chauvinists). In conjunction, these two forces influenced all things and were present individually or together in all physical objects and situations. In the case of numbers, odd numbers were Yang and even, Yin. For a harmonious state of being to exist, Yin-Yang forces had to be balanced. Through this concept, a cosmological numerology was devised. This numerology came alive in the Lo shu.

Now let us examine the Yin-Yang principle in action in the conception and construction of the Lo shu (fig. 2). In step c, we exchanged Yang elements; but a magic square of order three could also have been achieved by an exchange of Yin elements. In observing the mathematical properties of this square, we find that the sum of every row, column, and diagonal is 15. The product of the central entry C and the order of the square R equals the constant sum S, that is, $R \times C = S$; further, $R^2 \times C = T$, the total sum of all elements, in this case 45. These formulas will work for all odd-order squares.

Obviously, the ancient Chinese knew and used these formulas in devising their squares. The central entry 5, which is located between opposing pairs, is also their arithmetic mean; i.e.,

$$\frac{8+2}{2} = 5, \frac{6+4}{2} = 5,$$

and so on. The number 5 emerges as a symbol of centrality. The Lo shu configuration is kept in balance, or harmony, by the centrality of 5, and the whole configuration was considered a symbol of universal harmony. Centrality was a revered concept in ancient Chinese thought. In viewing the square, a Han scholar could see the Five Directions of old China: 1, north; 9, south; 3, east; 7, west; and 5, the center. The square itself became a symbolic representation of ancient China through an appropriate symbolic interpretation of its elements—Yang numbers were associated with firm and strong objects, and thus the Yang numbers of the square represented the Five Sacred Mountains of Chinese history; Yin numbers, being sinuous and yielding, represented the four major river systems. By performing permutations on the Lo shu, the ancient geomancer could also represent the cycles of nature—the four seasons. The Lo shu was the most potent and revered of all the magic squares

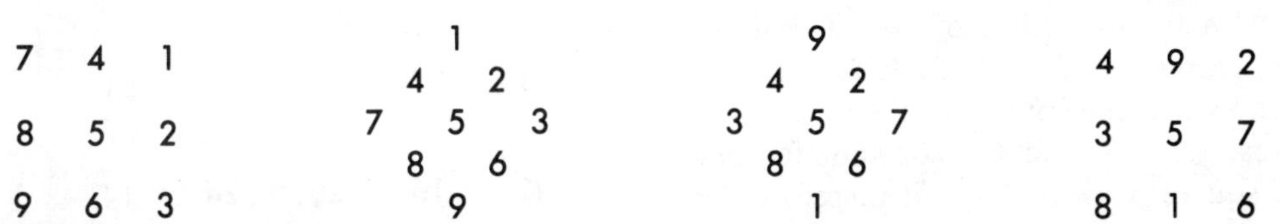

(a) Construct a natural square of order three.

(b) Distort the square into a diamond configuration.

(c) Exchange the opposing Yang elements.

(d) Compress into a square.

Fig. 2

21	16	11	6	1
22	17	12	7	2
23	18	13	8	3
24	19	14	9	4
25	20	15	10	5

(a) Construct a natural square of order five.

		7		
	12		8	
17		13		9
	18		14	
		19		

(b) Distort the central core.

		7		
	14		18	
17		13		9
	8		12	
		19		

(c) Apply Yin technique on the elements.

14	7	18
17	13	9
8	19	12

(d) Compress into a square of order three.

1	23	16	4	21
15	14	7	18	11
24	17	13	9	2
20	8	19	12	6
5	3	10	22	25

(e) Arrange the outer border to form a magic square of order five.

Fig. 3

of old China. Other squares were constructed along Lo shu principles.

The magic square of order five is constructed in figure 3. This process is mechanical with the exception of step e, but remember, by use of the formula $R \times C = S$ we know that the constant sum must be 65, that is, 13×5, and the constant for our core square is $13 \times 3 = 39$; therefore the border will consist of opposing Yin or Yang number pairs whose sum is $65 - 39 = 26$.

A still more harmonious and mathematically mysterious magic square of order five was employed by the Chinese (fig. 4).

12	27	33	23	10
28	18	13	26	20
11	15	21	17	31
22	16	29	24	14
32	19	9	15	30

Fig. 4

Apparently this square was devised from the natural sequence 1–25 by the addition of 8 to each entry; thus, in analyzing it, we must first reduce each entry by 8 to obtain figure 5.

④	19	㉕	15	②
20	10	5	18	12
③	17	⑬	9	㉓
14	8	21	16	6
㉔	11	①	7	㉒

Fig. 5

If we perform repeated subtractions of 8 from the circled entries in figure 5 until a one-digit number remains in each position, we can see that the square has a Lo shu framework (fig. 1). The remaining numbers are deployed in a balanced fashion in complementary Yin or Yang pairs about the central entry, 13: the sum of each pair equals 26—twice the central entry—and they are positioned symmetrically about it. In the former magic square of order five, the sum of complementary pairs also equaled 26. If the digits for the central entry are added together, their sum is 4; if the digits of 26 are added together, their sum is 8, or 2 × 4. "Fourness," then, becomes a fundamental property of these squares; 4, however, a Yin number for the Chinese, represents the earth and is inauspicious in itself unless balanced by the Yang number 3, which represents heaven. By adding 8 to all entries of the second square, the ancient Chinese alleviated this undesirable situation.

Historically, magic squares of order six have been difficult to devise. The Chinese used the Lo shu principle to construct two such squares (figs. 6 and 7). The method of construction appears similar for both (fig. 8).

13	22	18	27	11	20
31	4	36	9	29	2
12	21	14	23	16	25
30	3	5	32	34	7
17	26	10	19	15	24
8	35	28	1	6	33

Fig. 6

4	13	36	27	29	2
22	31	18	9	11	20
3	21	33	32	25	7
30	12	5	14	16	34
17	26	19	28	6	15
35	8	10	1	24	33

Fig. 7

Now in order to make the square in figure 8c a proper magic square, elements in the central and bottom corner subsquares had to be rearranged, thus giving us the first square of order six (fig. 6). The second order-six square (fig. 7) is also constructed from step c, but the exact method employed is not clear. An amazing property of these squares is that if the digits of each entry are added together, all subsquare entries are reduced to their Lo shu elements, and then if one sums these entries, the result is a new magic square of order three (fig. 9).

Further, the total sum of each of these order-six squares just considered is 666—the demonic symbol of medieval European numerology! In turn, when these digits are

28	19	10	1
29	20	11	2
30	21	12	3
31	22	13	4
32	23	14	5
33	24	15	6
34	25	16	7
35	26	17	8
36	27	18	9

(a) Write the natural sequence 1–36 in four columns (note sum of digits in each entry).

28	19	10	1→10	19
			28	1

(b) Rewrite each row as a 2 X 2 matrix by listing consecutive entries as diagonal opposites (demonstrated with row 1).

13	22	18	27	11	20
31	④	36	⑨	29	②
12	21	14	23	16	25
30	③	32	⑤	34	⑦
17	26	10	19	15	24
35	⑧	28	①	33	⑥

(c) Order squares according to the correspondence of their bottom right entries with the entries of the Lo shu in figure 2d.

Fig. 8

4	4	9	9	2	2
4	4	9	9	2	2
3	3	5	5	7	7
3	3	5	5	7	7
8	8	1	1	6	6
8	8	1	1	6	6

(a) Add digits of each entry until a one-digit result remains.

16	36	8
12	20	28
32	4	24

(b) Summing entries of subsquares, we obtain a magic square of order three.

Fig. 9

added together, their sum reduces to 9, as did the total sum of the original Lo shu, that is, 4 + 5 = 9, and as will the sum of the digits of the total sum for the magic square of order nine yet to be considered. This phenomenon further illustrates the intrinsic kinship of the system of magic squares devised by the Chinese.

The magic square of order seven also used Lo shu principles in its construction (fig. 10). The remaining entries were then arranged about this core according to previously discussed principles.

Perhaps the most impressive magic square of old China, in terms of its size and simplicity of construction, was the square of order nine. The construction is shown in figure 11. Now, positionally order the resulting nine squares of order three by the correspondence of their bottom central entry with those of the Lo shu, that is, 4, 9, 2; 3, 5, 7; 8, 1, 6. The resulting composite magic square is our desired one of order nine (fig. 12).

Conclusion

This discussion of ancient Chinese magic squares and their properties has been far

1	8	15	22	29	36	43
2	9	16	23	30	37	44
3	10	17	24	31	38	45
4	11	18	25	32	39	46
5	12	19	26	33	40	47
6	13	20	27	34	41	48
7	14	21	28	35	42	49

(a) Construct a natural square of order seven.

		23		
	17		31	
11		25		39
	19		33	
		27		

(b) Select a diamond-shaped core of central entries.

		23		
	33		19	
11		25		39
	31		17	
		27		

(c) Rearrange the entries according to Yin technique.

33	23	19
11	25	39
31	27	17

(d) Compress them into a magic square of order three (note sum of digits for each entry).

Fig. 10

1	10	19	28	37	46	55	64	73
2								
3								
4								
5								
6								
7								
8								
9								81

(a) Construct a natural square of order nine.

55	28	1
64	37	10
73	46	19

(b) Fold each row into a square of order three (demonstrated with row one).

		1		
	28		10	
55		37		19
	64		46	
		73		

(c) Distort it into a diamond configuration.

		73		
	28		10	
19		37		55
	64		46	
		1		

(d) Apply the Yang exchange principle.

28	73	10
19	37	55
64	1	44

(e) Compress it into an order-three square.

Fig. 11

31	76	13	36	81	18	29	74	11
22	40	58	27	45	63	20	38	56
67	④	49	72	⑨	54	65	②	47
30	75	12	32	77	14	34	79	16
21	39	57	23	41	59	25	43	61
66	③	48	68	⑤	50	70	⑦	52
35	80	17	28	73	10	33	78	15
26	44	62	19	37	55	24	42	60
71	⑧	53	64	①	46	69	⑥	51

Fig. 12

from inclusive either in scope or in the exploration of the specific squares considered. A more thorough exploration of the squares above is left for the readers and their students. The magic square of order nine will provide in itself a cornucopia of fascinating number patterns and phenomena for the inquisitive investigator to discover.

In 1275 A.D., the scholar-historian Yang Hui wrote a compendium on early Chinese mathematics, *Hsu-ku chai-chi suan fa* [Continuation of ancient mathematical methods for elucidating the strange properties of numbers], in which he discussed the above-mentioned magic squares and many others his predecessors had devised. From the evidence of this book and other records, it appears that the contributions of early China to the science and art of constructing magic squares was quite significant. The Chinese invented the magic square of order three; they also devised the first magic squares of order five, six, seven, and nine, bordered magic squares, and composite and augmented squares. At the heart of all these innovations lies either the mysterious Lo shu itself or adaptations of it. The Lo shu emerges from the history of science as an example in which philosophy, religion, and mathematics were combined in a complex creation of human thought—a two-thousand-year-old creation that can still intrigue and stimulate its beholder with its "magical" properties.

REFERENCES

Camman, Schyler. "Old Chinese Magic Squares." *Sinologica* 7 (1962):14–53.

Needlham, Joseph. *Science and Civilization in China*, vol. 3. Cambridge: At the University Press, 1955.

MINICALCULATORS AND REPEATING DECIMALS

Here we go again on a topic that continues to interest teachers and students. This time, with invaluable insights from our calculators . . . Here we go again on a topic . . .

By DONOVAN R. LICHTENBERG

University of South Florida
Tampa, FL 33620

In recent years this journal has published several articles concerned with the theory of repeating decimals (see Hutchinson 1972, Alexander 1974, Anderson 1974, and Jacobs 1975). As with many mathematical topics, this theory is usually easier for young students to comprehend if specific examples can be verified by computation. But most students do not have the fortitude to carry out tedious pencil-and-paper computation, and consequently they have not studied decimals very carefully. The purpose of this article is to show how an inexpensive minicalculator can be used advantageously to determine repeating decimals. It will be seen that even the decimal representations of numbers such as 1/17 and 1/19 can be easily computed. We will cover just enough of the theoretical aspects of the subject to make the discussion meaningful. It is to be hoped that with minicalculators to do most of the computation, students will delve into this topic to a greater extent than they have in the past. It is a fascinating subject that may lead to a rewarding study of number theory.

It will be assumed in what follows that we are working with a minicalculator that has an eight-digit display and does not round off. (You can easily make the necessary adjustments if your calculator differs.) If a student uses such a calculator to determine decimal equivalents for 1/2, 1/3, 1/4, 1/5, 1/6, 1/7, 1/8, and 1/9 in succession, the calculator will display the following:

0.5
0.3333333
0.25
0.2
0.1666666
0.1428571
0.125
0.1111111

It appears that 1/2, 1/4, 1/5, and 1/8 have terminating decimal representations whereas the others do not. (Of course, it is possible that the others terminate after seven or more digits.)

To determine precisely the rational numbers that can be represented by terminating decimals, we make use of the fact that any terminating decimal is equivalent to a fraction whose denominator is a power of ten. For example,

$$0.142 = \frac{142}{1000} = \frac{142}{10^3} .$$

This particular fraction is not in lowest terms (i.e., the numerator and denominator are not relatively prime), because $142 = 2 \cdot 71$ and $10^3 = 2^3 \cdot 5^3$. It follows that

$$0.142 = \frac{2 \cdot 71}{2^3 \cdot 5^3} = \frac{71}{2^2 \cdot 5^3} = \frac{71}{500} .$$

The important thing to recognize is that the denominator has no prime factors other than 2 or 5. Since $10^n = 2^n \cdot 5^n$ for any natural number n, the denominator of the fraction in lowest terms for a given terminating decimal can not have prime factors other than 2 or 5. Conversely, if a/b is in lowest terms and b has no prime factor other than 2 or 5, it is possible to find an equivalent fraction whose denominator is a power of ten, and this fraction results in a terminating decimal. The following examples should serve to illustrate this:

$$\frac{1}{2} = \frac{1 \cdot 5}{2 \cdot 5} = \frac{5}{10} = 0.5$$

$$\frac{1}{4} = \frac{1}{2^2} = \frac{1 \cdot 5^2}{2^2 \cdot 5^2} = \frac{25}{100} = 0.25$$

$$\frac{1}{5} = \frac{1 \cdot 2}{5 \cdot 2} = \frac{2}{10} = 0.2$$

$$\frac{1}{8} = \frac{1}{2^3} = \frac{1 \cdot 5^3}{2^3 \cdot 5^3} = \frac{125}{1000} = 0.125$$

$$\frac{1}{20} = \frac{1}{2^2 \cdot 5} = \frac{1 \cdot 5}{2^2 \cdot 5^2} = \frac{5}{100} = 0.05$$

$$\frac{1}{25} = \frac{1}{5^2} = \frac{1 \cdot 2^2}{5^2 \cdot 2^2} = \frac{4}{100} = 0.04$$

We conclude that if a/b is in lowest terms, there is a terminating decimal representation if and only if b contains no prime factors other than 2 or 5.

It is possible to show that a rational number can be represented either by a terminating decimal or by an infinite repeating decimal. The easiest explanation for students to understand involves an examination of their work after they have determined a decimal for 1/7 by means of the usual division algorithm. The circled remainders in figure 1 should be noted.

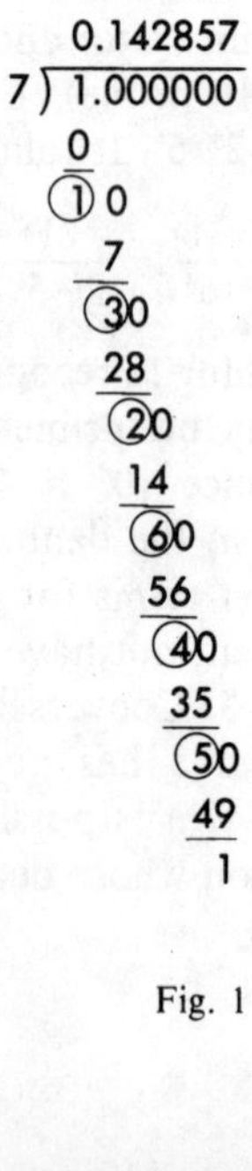

Fig. 1

We know from our previous analysis that there is no terminating decimal for 1/7. Hence, we will not arrive at a remainder of 0 in the division process. The only possible remainders when a number is divided by 7 are 1, 2, 3, 4, 5, and 6. The circles in figure 1 indicate that each of these remainders has occurred once. Therefore, the next remainder must be the same as a previous one, and a repeating decimal results. The rational number 1/7 can be represented by the repeating decimal

$$0.142857142857142857\ldots,$$

where the block of six digits repeats infinitely many times. This block of digits is called the *repetend* (also *period* or *cycle*). The standard way of denoting a repeating decimal is to place a bar over the repetend:

$$\frac{1}{7} = 0.\overline{142857}$$

In general, suppose that $a < b$ and that we have concluded that there is no terminating decimal for a/b. We then know that we will not arrive at a remainder of 0 when applying the division algorithm. Hence, the only numbers that can possibly occur as remainders are $1, 2, 3, \ldots, b - 1$. If all the numbers occur, the next remainder must duplicate a previous one. In some cases a duplicate will occur before all the possible remainders have been used. Either way, a repeating decimal results. It can be seen that the maximum number of digits possible in the repetend is $b - 1$.

We could at this point get into a lengthy discussion about the meaning of an infinite decimal expression and the justification of the algorithm used to obtain it. But such a discussion would represent too much of a detour for the purposes of this article. After a certain amount of experience with decimals, students usually are willing to accept an infinite decimal as an expression for a number. They also are willing to accept the extension of the division algorithm that they first learned when working with whole numbers.

It is worth noting that if the decimal point in figure 1 is ignored, the work is precisely the same as if one had divided 10^6 by 7. We can see that if 10^6 is divided by 7, we get a quotient of 142857 and a remainder of 1. That is, $10^6 = 142857 \cdot 7 + 1$. (Observe that 7 is a factor of $10^6 - 1$, a fact whose significance will be discussed later.)

Let's examine some other decimals using a minicalculator. If we divide 1 by 17, the calculator displays

$$0.0588235.$$

Since there are seven digits to the right of the decimal point, we know that in the corresponding pencil-and-paper computation it would have been necessary to annex seven zeros. It is an important fact that the calculator does not round off. We are thus sure of the first seven digits in the decimal expansion. If we knew what the remainder would be at this stage, it would be possible to carry the computation further. The computation thus far would be the same as if we were dividing 10^7 by 17. We know from the calculator display that if we did divide 10^7 by 17, the quotient would be 588235. The remainder would then be $10^7 - 588235 \cdot 17$. When we multiply the entry in the calculator by 17, we get 0.9999995, and we conclude that $588235 \cdot 17 = 9999995 = 10^7 - 5$. Hence, $10^7 = 588235 \cdot 17 + 5$. The remainder at this stage of the division is thus 5, and we can extend the computation by dividing 5 by 17. When we do so, the calculator displays

$$0.2941176.$$

We therefore know the next seven digits in the decimal for 1/17. That is, we know that

$$\frac{1}{17} = 0.05882352941176\ldots,$$

with perhaps some more digits yet to be determined. If the last entry displayed in the calculator is multiplied by 17, we get 4.9999992. Since this differs from 5 (remember we divided 5 by 17) by .0000008, we know that 8 would be the remainder at this stage of the computation. We therefore continue by dividing 8 by 17. Note first that we have already determined fourteen digits of the decimal for 1/17 and we know that it has to repeat after no more than sixteen digits. When we divide 8 by 17, we obtain 0.4705882, and we see that the decimal is repeating. We have thus determined that

$$\frac{1}{17} = 0.\overline{0588235294117647}.$$

Let's try 1/19. If we divide 1 by 19, we obtain 0.0526315. Multiplying this by 19 gives us 0.9999985, so we know that the remainder must be 15 at this stage. Dividing 15 by 19, we get 0.7894736, and we know that

$$\frac{1}{19} = 0.05263157894736\ldots,$$

When the entry in the calculator is multiplied by 19, the calculator displays 14.999998. Here we must be careful. Note that because there are two digits to the left of the decimal point, there are only six to the right. We know there should be seven. The calculator has cut off the last digit. What is the missing digit? Since $6 \cdot 9 = 54$, the right hand digit in the product of 19 and 0.7894736 is 4. Thus the product is 14.9999984, and the remainder at this stage must be 16. Dividing this by 19, we obtain 0.8421052, and we conclude that

$$\frac{1}{19} = 0.\overline{052631578947368421}.$$

On examination of the decimal representations of 1/7, 1/17, and 1/19, a student might conclude that if the denominator is a prime p (different from 2 or 5), the length of the repetend is $p - 1$. However, the decimal for 1/3 shows that this is an incorrect conjecture. Using the minicalculator to determine the decimal for 1/13, we find another counterexample. When we divide 1 by 13, the display shows 0.076923. Note that there are only six digits to the right of the decimal point. When this is multiplied by 13, the calculator gives us 0.999999. We conclude that the remainder at this stage is 1 and that the decimal repeats after six digits:

$$\frac{1}{13} = 0.\overline{076923}$$

It can be seen that the seventh digit to the

right of the decimal point is 0. This is why the calculator didn't show it.

With the aid of a calculator, your students should be able to produce many examples to illustrate the theorems that follow.

THEOREM 1. *The length of the repetend for a fraction in lowest terms depends only on the denominator and not on the numerator.* (For a proof, see Ore 1948, p. 320.)

If a minicalculator is used to determine the decimal for 2/7, it would be found that

$$\frac{2}{7} = 0.\overline{285741}.$$

If students compare this with the decimal for 1/7, they would notice that the digits in the repetend are the same and in the same cyclical order; only the starting point is different. Students can see why if they examine the pencil-and-paper computations for 1/7 and 2/7. The arrows in figure 2 mark corresponding stages, after which the steps are identical. Consequently, the same digits

```
       0.142857              0.285714
    7 ) 1.000000          7 ) 2.000000
        0                     0
        10              -->   20
         7                    14
         30                    60
         28                    56
   -->    20                    40
          14                    35
           60                    50
           56                    49
            40                    10
            35                     7
             50                    30
             49                    28
              1                     2
```

Fig. 2

must appear in the quotient of each and in the same order. Moreover, just by looking at the computation for 1/7 we can conclude that

$$\frac{3}{7} = 0.\overline{428571},$$

$$\frac{4}{7} = 0.\overline{571428},$$

$$\frac{5}{7} = 0.\overline{714285},$$

and

$$\frac{6}{7} = 0.\overline{857142}.$$

THEOREM 2. *If the denominator of a fraction in lowest terms is of the form $2^r \cdot 5^s \cdot k$ where $k > 1$ and k does not contain 2 or 5 as a factor, then the number of nonrepeating digits in the decimal will be the maximum of r and s, and the number of digits in the repetend will be the least positive integer n such that k is a factor of $10^n - 1$.*

Note that according to this theorem there will be no nonrepeating digits if $r = s = 0$, that is, if the denominator has neither 2 nor 5 as a factor. In this case, we say that the decimal is a pure repeating decimal. (For a proof of this theorem, see Hutchinson 1972.)

With a calculator, a student would be able to determine that

$$\frac{1}{28} = 0.03\overline{571428}.$$

In applying theorem 2 to 1/28, $r = 2$, $s = 0$, and $k = 7$, because $28 = 2^2 \cdot 7$. Hence, the decimal should contain two nonrepeating digits, and the repetend should contain six digits because 7 is a factor of $10^6 - 1$ (and 7 is not a factor of $10^n - 1$ for $0 < n < 6$). Perhaps the reason for the two nonrepeating digits becomes more apparent if the decimal for 1/28 is determined as follows:

Since $\frac{1}{28} = \frac{1}{2^2 \cdot 7}$.

$$100 \cdot \frac{1}{28} = 2^2 \cdot 5^2 \cdot \frac{1}{2^2 \cdot 7}$$

$$= \frac{25}{7} = 3\frac{4}{7} = 3.\overline{571428}.$$

If $100 \cdot \frac{1}{28} = 3.\overline{571428}$,

then $\frac{1}{28} = \frac{3.\overline{571428}}{100} = 0.03\overline{571428}$.

It should be clear that we multiplied 1/28 by 100 because this results in a fraction whose denominator has neither 2 nor 5 as a factor. It can be seen that 100 is the least power of ten that will accomplish this. Note also that division by 100 in the last step accounts for the two nonrepeating digits and that the repetend depends only on the factor 7 in the denominator of 1/28.

Because of theorem 2, it is possible to determine fractions with relatively short repetends by examining the factorizations of numbers of the form $10^n - 1$ for small n. For n less than or equal to 6, it is fairly easy to determine the factorizations:

$$10^1 - 1 = 3^2$$
$$10^2 - 1 = 3^2 \cdot 11$$
$$10^3 - 1 = 3^3 \cdot 37$$
$$10^4 - 1 = 3^2 \cdot 11 \cdot 101$$
$$10^5 - 1 = 3^2 \cdot 41 \cdot 271$$
$$10^6 - 1 = 3^3 \cdot 7 \cdot 11 \cdot 13 \cdot 37$$

(A calculator, of course, can be of assistance in determining such factorizations. Students know that $a - 1$ is a factor of $a^n - 1$, which implies that $10 - 1$ or 3^2 is a factor of $10^n - 1$. This provides a way to begin, and the calculator can help with the rest.)

From the list above it can be seen, for example, that a fraction with a denominator of 101 will have a repetend with four digits because 101 is a factor of $10^4 - 1$ and not a factor of $10^n - 1$ for any smaller n. Similarly, a denominator of 41 will result in a repetend of five digits, and so forth. The minicalculator can be used to verify that

$$\frac{1}{101} = 0.\overline{0099},$$

and

$$\frac{1}{41} = 0.\overline{02439}.$$

The calculator will also show that

$$\frac{1}{11} = 0.\overline{09},$$

$$\frac{1}{37} = 0.\overline{027},$$

and

$$\frac{1}{271} = 0.\overline{00369}.$$

These results all agree with the factorizations given above.

We can also see from the list of factorizations that 21 is a factor of $10^6 - 1$ because $21 = 3 \cdot 7$. Hence, a fraction (in lowest terms) whose denominator is 21 will have a repetend of 6 digits. The calculator will show that

$$\frac{1}{21} = 0.\overline{047619}.$$

As was observed earlier, the maximum number of digits in the repetend of a/b is $b - 1$. Students will note that in some cases (e.g., denominators of 7, 17, or 19) the number of digits in the repetend is the maximum possible. In other cases (e.g., denominators of 3, 11, or 13), it is not. An alert student may note that whenever the denominator is a prime p, the number of digits in the repetend is a factor of $p - 1$. This is a corollary of the following theorem.

THEOREM 3. *If a/b is in lowest terms and b contains neither 2 nor 5 as a factor, then the number of digits in the repetend is a factor of $\phi(b)$ where $\phi(b)$ is the number of positive integers less than b that are relatively prime to b.* (For a proof, see Rademacher and Toeplitz, pp. 153–55.)

For a prime p, it can be seen that $\phi(p) = p - 1$, because every positive integer less than p is relatively prime to p. For a composite number n, $\phi(n)$ is always less than $n - 1$. The numbers less than 21 and relatively prime to 21 are 1, 2, 4, 5, 8, 10, 11, 13, 16, 17, 19, and 20. Thus, $\phi(21) = 12$. As theorem 3 predicts, the number of digits in the repetend for 1/21 is a factor of $\phi(21)$.

Now let's examine the decimal for 2/21. The calculator shows that

$$\frac{2}{21} = 0.\overline{095238}.$$

Comparing this with the decimal for 1/21, we find that the repetends do not appear to be related. Further computation with the calculator would reveal the following:

$$\frac{1}{21} = 0.\overline{047619}$$

$$\frac{2}{21} = 0.\overline{095238}$$

$$\frac{4}{21} = 0.\overline{190476}$$

$$\frac{5}{21} = 0.\overline{238095}$$

$$\frac{8}{21} = 0.\overline{380952}$$

$$\frac{10}{21} = 0.\overline{476190}$$

$$\frac{11}{21} = 0.\overline{523809}$$

$$\frac{13}{21} = 0.\overline{619047}$$

$$\frac{16}{21} = 0.\overline{761904}$$

$$\frac{17}{21} = 0.\overline{809523}$$

$$\frac{19}{21} = 0.\overline{904761}$$

$$\frac{20}{21} = 0.\overline{952380}$$

Note that there are twelve fractions in lowest terms with denominator 21 and numerator less than 21 because $\phi(21) = 12$. Inspection of the list above shows that the repetends for 1/21, 4/21, 10/21, 13/21, 16/21, and 19/21 have the same digits in the same cyclical order and that the repetends for 2/21, 5/21, 8/21, 11/21, 17/21, and 20/21 are similarly related. Pencil-and-paper computation would show that when 1 is divided by 21, only the numbers 1, 4, 10, 13, 16, and 19 occur as remainders. When 2 is divided by 21, the other six integers that are less than 21 and relatively prime to it occur as remainders. (Why won't a number that has a factor in common with 21 show up as a remainder?)

The partitioning of the set of fractions (in lowest terms) with a given denominator can be further illustrated by an examination of fractions with denominator 41. We saw earlier that $1/41 = 0.\overline{02439}$. If the minicalculator is also used to determine decimals for 2/41, 3/41, 4/41, . . . , 40/41, it would be apparent that the fractions fall into eight sets of five according to their repetends. In the set with 1/41 we would find the following:

$$\frac{10}{41} = 0.\overline{24390}$$

$$\frac{16}{41} = 0.\overline{39024}$$

$$\frac{18}{41} = 0.\overline{43902}$$

$$\frac{37}{41} = 0.\overline{90243}$$

Another set would consist of the following:

$$\frac{2}{41} = 0.\overline{04878}$$

$$\frac{20}{41} = 0.\overline{48780}$$

$$\frac{32}{41} = 0.\overline{78048}$$

$$\frac{33}{41} = 0.\overline{80487}$$

$$\frac{36}{41} = 0.\overline{87804}$$

The interested reader can determine the other six sets.

At the beginning of this article we mentioned that this topic may stimulate interest in number theory. Of historical importance is the fact that this is exactly the topic that provided so much motivation for Carl Friedrich Gauss when he was in his teens. His investigations ultimately led to the publication of his *Disquisitiones Arithmeticae* (Researches in Arithmetic) when Gauss was twenty-four. This book, considered the beginning of modern number theory, contains everything that we have discussed here and much, much more. Gauss in his teens had computed the decimals for the reciprocals of all integers up to 1000. And he, of course, didn't have a minicalculator. With one, your students can taste some of the wonder and amazement Gauss did, with less tedious work.

BIBLIOGRAPHY

Alexander, F. Doyle. " 'One Small Jump'—into Repeating Decimals and Prime Numbers." *Mathematics Teacher* 67 (October 1974):520–25.

Anderson, John T. "Periodic Decimals." *Mathematics Teacher* 67 (October 1974):504–09.

Bell, Eric T. *Men of Mathematics.* New York: Simon and Schuster, 1937.

Gauss, Carl F. *Disquisitiones Arithmeticae.* Translated by Arthur A. Clarke. New Haven: Yale University Press, 1966.

Hutchinson, Margaret R. "Investigating the Nature of Periodic Decimals." *Mathematics Teacher* 65 (April 1972):325–27.

Jacobs, Neal. "More on Repeating Decimals." *Mathematics Teacher* 68 (March 1975):249–52.

Ore, Oystein. *Number Theory and Its History.* New York: McGraw-Hill Book Co., 1948.

Rademacher, Hans, and Otto Toeplitz. *The Enjoyment of Mathematics.* Princeton: Princeton University Press, 1957.

Woodburn, Douglas. "Can You Predict the Repetend?" *Mathematics Teacher* 69 (December 1976):675–79.

Answers

("Calculating Order," p. 142)

Where the last digit is underlined the number has been rounded.

Sheet 1: (1) 63, 255; (2) 88, 40; (3) 217.5, 14.0625; (4) 233, 233; (5) 59, 59; (6) 193.5, 193.5; (7) 185, 185; (8) $355.666\underline{7}$, 99; (9) $13.714\underline{3}$, $107.761\underline{9}$.

Sheet 2: (1) 30.758, 68.3032, (2.2 × 3.71 + 4.035)5.6; (2) 22.244, 30.236, (2.16 + 3.02)4.7 + 5.89; (3) $-1.687\underline{3}$, $-2.440\underline{2}$, (5.7/3.8 − 6.4)/2.008; (4) 78.58, 33.22, (7.2 × 8.4 + 5.6)/4 + 16.7; (5) $39.652\underline{9}$, $75.970\underline{6}$, (4.9/5.1 × 6.3 + 4.8)7; (6) $34.251\underline{4}$, $3.892\underline{9}$, ((3.28 × 7.1 + 4.26)2.5 + 6.03)/3.7/5.2; (7) $23.023\underline{5}$, $1.196\underline{6}$, (4.26 × 5.9 + 9.8)/7.6 − 3.4; (8) $5.864\underline{8}$, $-0.199\underline{4}$, ((5.08 + 7.62)/6.8 − 4.6)/13.7.

Sheet 3: (1) (7.8 + 6.7)21.4 + 4.7, 315; (2) (32.6(25.9) + 19.8 − 7.5)/2, 428.32; (3) ((9.02 + 2.48)/3.4 + 5.4)7.3, 64.1112; (4) $(7.03 + 4.66)^2 \times 6.42$, $877.332\underline{2}$; (5) ((4.3/7.4 + 3.7)6.9 + 8.6)2.3, $87.720\underline{8}$; (6) (4(3.1)(4.2) + 16.2)/7.8, $8.753\underline{8}$; (7) ((4.2 + 9.2)5.2/6 + 7.9)3.6/5.3, $13.254\underline{3}$; (8) (((3.5 + 5.3)4.2 + 17.4)6.7 − 7.6)/3 − 8.8, $110.070\underline{7}$; (9) ((7.6 + 8.1)3.02/8.62 + 5.1)/3.4/4.2, $0.742\underline{3}$.

THE POCKET CALCULATOR AS A TEACHING AID

Encourage your students to use their minicalculators to discover patterns and verify mathematical statements.

By ELI MAOR

University of Wisconsin—Eau Claire
Eau Claire, WI 54701

SINCE its first appearance in the market about four years ago, the pocket calculator has quickly become everyone's home tool, as common as the transistor radio or the cassette tape recorder. The increasing variety of calculators, combined with their generally declining prices, has made them available to more and more people. As a result, even people who previously had no interest in mathematics have begun to express an interest in the properties of numbers, the operations with them, and the relations that exist among them.

The stimulation of exploring with these calculators apparently has a psychological effect on people's attitude toward numbers and arithmetic. The boring, time-consuming, and ever-dreaded use of tables—whether they are the ordinary multiplication tables of elementary school days or the logarithmic and trigonometric tables of high school—suddenly disappears. Instead, one plays with a small, pocket-sized, attractive instrument, operated by the touch of a few keys, and generates answers that appear electronically in the display.

Undoubtedly, the introduction of the pocket calculator will bring about a drastic change in the field of elementary computations, that is, computations that are too trivial to be processed by a computer, yet too time-consuming for manual calculation. This trend will have an obvious bearing on the teaching of mathematics in schools. Not only will the use of most numerical tables become unnecessary, but the teaching of many topics in elementary as well as college mathematics will be given new insight, interest, and even fun by the use of these instruments. It is the purpose of this article to offer several examples and suggestions on how this might be accomplished.

I should like to thank Professor Henry Mullish, of the Courant Institute of Mathematical Sciences, who inspired in me the interest and enthusiasm for using the pocket calculator as an educational tool, and without whose ideas and help this article would never have been written.

Arithmetic and Algebra

The proofs of all the basic formulas and identities of elementary algebra can, and should, be supplemented by testing their validity for various numerical values of the symbols appearing in them. This is where the pocket calculator is of great help. Not only will it increase students' faith in the formula, but it will also encourage them to test and search for new relations and patterns.

As a first step, however, some simple experiments should be performed in class in order to test the calculator's logic in processing sequences of operations. Calculators may be classified into two categories according to their algebraic hierarchy, that is, according to the method used to process sums of products. In the one category, an expression such as $ab + cd$ is computed exactly as we would understand it, for example, $1 \times 2 + 3 \times 4 = 2 + 12 = 14$. In the other category, the calculator will interpret the same expression as $(ab + c) \times d$, that is, $(1 \times 2 + 3) \times 4 = 20$. Generally, the most sophisticated models, such as Texas Instruments' SR-50, belong to the first category, and the simpler types, such as Texas Instruments' SR-10, belong to the second.

A similar situation exists with respect to sums of fractions, sums of squares, sums of roots, and so on. It is very important to be aware of these differences before a class goes on to more complex calculations. The interpretations of several expressions for two categories of calculators can best be summarized in table 1. The last column in table 1 indicates the ways the expressions in column one should be interpreted for calculators in the second category.

TABLE 1
TWO INTERPRETATIONS FOR SOME SPECIFIC ALGEBRAIC EXPRESSIONS

Expression	Interpretation Category 1	Interpretation Category 2	Correct Input Notation for Category 2
$ab + cd$	$ab + cd$	$(ab + c)\,d$	$\left(\frac{ab}{d} + c\right) d$
$\frac{a}{b} + \frac{c}{d}$	$\frac{a}{b} + \frac{c}{d}$	$\frac{\frac{a}{b} + c}{d}$	$\left(\frac{ad}{b} + c\right) / d$
$a + b^2$	$a + b^2$	$(a + b)^2$	$b^2 + a$
ab^2	ab^2	$(ab)^2$	$b^2 a$

A discussion can now be initiated on these differences, and the students can practice with their own calculators to discover their various features.

The explorations with some of the more sophisticated operations of the calculator might turn into exciting discoveries for the beginner. Let us take the reciprocal key as an example. A simple formula like $\frac{1}{1/x} = x$, with which many students are familiar only as a formal means of handling composite fractions, is suddenly given a very realistic meaning. Choose any number (except zero, of course), press the "$1/x$" key twice, and the very same number appears in the display! The discovery can be made even more exciting by asking the students to watch the results as they press the "$1/x$" key first an even number of times and then an odd number of times. Almost without noticing it, we have introduced the students to the transformation equations $T^{2n} = I$ and $T^{2n+1} = T^{-1}$, which characterize all involutory transformations such as $y = 1/x$.

Similar experiments can be performed with the square and square-root keys. On some instruments, such as the SR-50, any successive application of two inverse functions will restore the exact original number; whereas in other types, such as the SR-10, the result will be only approximately equal to the original number. In this latter case, it will be interesting to observe what happens when a sequence of x^2 and $\sqrt{x}$ operations is performed on the same number x. Does the error remain bounded, or does it accumulate?

It is also fun to test the calculator's reaction to "forbidden" operations, such as zero division. Some types will protest with a violent flash of the display; other, shyer types will merely indicate with an error sign that the operation is invalid. An interesting situation is exhibited on the SR-50: if you take the common logarithm of zero, you get, as expected, the largest negative number the display can show, −9.999999999 times 10^{99}, which is the instrument's interpretation of minus infinity. If, however, you take the *natural* logarithm of zero, you get the unexpected flashing result −4.342944819 times 10^{99}! Each type of minicalculator has its own idiosyncracies.

Finally, many interesting and quite surprising numerical relations can be discovered with the calculator. Who would guess, for example, that e^π and π^e have almost the same value? ($e^\pi = 23.14069\ldots$; $\pi^e = 22.45915\ldots$). Or that $2 - \sqrt{3}$ is exactly the reciprocal of $2 + \sqrt{3}$? Of course, after discovering these relations on the calculator, the class can go on to analyze them and find out their origin. Thus, $e^\pi \sim \pi^e$ because the value of π is very nearly equal to that of $e \ln \pi$, so that $\pi^e \equiv e^{e \ln \pi} \sim e^\pi$. In the second example we have, of course, $(2 - \sqrt{3})(2 + \sqrt{3}) = 4 - 3 = 1$; incidentally, $2 - \sqrt{3}$ is the tangent of 15°, and $2 + \sqrt{3}$ is the tangent of 75°, and our relation becomes the trigonometric equation $\tan 15° \times \tan 75° = 1$. There is almost no limit to the ingenuity of numerical relations and properties that can be discovered with the calculator, and it can turn many a boring mathematics lesson into an exciting and unforgettable experience.

Trigonometry

There are obvious advantages in using the calculator over the conventional trigonometric tables: greater speed and accuracy and the lack of the need to calculate proportional parts. Most important, it is not necessary to reduce any angle to the range $0° \leq \alpha \leq 90°$, for the calculator will do this automatically. Anyone who has ever worked with angles whose measures are greater than 90° knows how time-consuming these reductions can be. These advantages, however, are somewhat counterbalanced by the fact that all angles must be entered either in radians or in decimal form, not in degrees/minutes/seconds. Only in some very advanced models, such as Texas Instruments' SR-51, is there direct conversion from decimal parts of a degree to minutes and seconds.

The pocket calculator can be of great help in demonstrating the validity of trigonometric identities. It is one thing to prove an identity but quite another to grasp the nature of the formula under consideration. Has anyone once bothered to test in the classroom the validity of, say, the formula $\sin^2\alpha + \cos^2\alpha = 1$ for various values of α, not just for 30° or 45°? Most probably, the answer is no. With the calculator, such a test becomes real fun (table 2). Now it depends on the specific type of calculator being used whether the result, after the "=" key has been pressed, is exactly 1 or 0.999999 . . . . The latter result is due to the calculator's truncating property, a point that should be explained to the class.

TABLE 2
VERIFYING $\sin^2\alpha + \cos^2\alpha = 1$

Choose any number for α	
Keyboard	Display
α	α
sin	$\sin\alpha$
x^2	$\sin^2\alpha$
+	$\sin^2\alpha$
α	α
cos	$\cos^\alpha$
x^2	$\cos^2\alpha$
=	

Another class of identities that can be demonstrated on the calculator are the trigonometric–inverse trigonometric relations. Identities such as $\sin(\arcsin x) = x$, $\arcsin(\sin x) = x$, or $\arcsin x = \arccos \sqrt{1 - x^2}$ appear quite abstract and meaningless to many students. A simple numerical demonstration will at once make these relations meaningful (table 3).

TABLE 3
VERIFYING $\arcsin(\sin\alpha) = \alpha$

Choose any number for α	
Keyboard	Display
α	α
sin	sin
arcsin	α

NOTE: The procedure of the operation "arcsin" differs from one instrument to another. Often it is required to press two keys, "arc" and "sin" or "inv" and "sin."

Finally, a very common error in the first lessons of a trigonometry course is to confuse, for example, $\sin 2x$ with $2 \sin x$. One or two demonstrations on the calculator will remove any notion that the two expressions are the same. Indeed, the correct relation, $\sin 2x = 2 \sin x \cos x$, can easily be demonstrated in the same way.

Functions and Limits

In the calculation of numerical values of functions, the pocket calculator is almost indispensable. Even simple functions, such as $y = ax^2 + bx + c$ or $y = \sin(ax+b)$, require quite involved computations, unless the constants a, b, c have integral values and only as long as we confine ourselves to integral values of x as well. But sometimes assigning only integral values to x will make the values of y too sparse for a graph to be meaningful. This is particularly true near a singular point of the function. For example, the function $y = \frac{2x + 1}{2x - 1}$ yields ever-increasing values as x approaches the value 0.5. It is imperative to calculate the values of y for x's very near to this point.

This is a task that fits exactly the capabilities of even the simplest calculator (table 4).

TABLE 4

EVALUATING $y = \dfrac{2x+1}{2x-1}$ FOR $x = 0.7$

	Keyboard	Display	Keyboard	Display	
denominator	0.7	0.7	0.7	0.7	numerator
	×	0.7	×	0.7	
	2	2	2	2	
	—	1.4	+	1.4	
	1	1	1	1	
	=	0.4	=	2.4	
	STO	0.4	÷	2.4	
			RCL	0.4	
			=	6.	

NOTE: If the calculator has no memory (STO key), one would have to write down the intermediate result, 0.4, manually and then use it again in the final division. In more advanced models, where there is an interchange function ("$x \rightleftarrows y$" key), one may calculate first the numerator, store it, then calculate the denominator, divide it into the result of the numerator, and then press the interchange key (before pressing the "=" key); or, alternatively, press the "=" key and then the reciprocal "$1/x$" key.

The same calculations should then be repeated for $x = 0.6, 0.55, 0.525$, and so on; then for $x = 0.3, 0.4, 0.45, 0.475$, and so on.

One may use similar procedures to show how the quotient $\dfrac{\Delta y}{\Delta x}$ of a continuous function $y = f(x)$ approaches the value of the derivative $y' = \dfrac{dy}{dx}$ as Δx becomes smaller and smaller near the point x_0 where the value of the derivative is sought.

Perhaps the most elegant use of the calculator as an educational tool is in the demonstration of limiting processes, such as the convergence of infinite sequences, series, and products. The calculator will not only calculate the values of the terms themselves but will automatically record the value of the partial sum (or product, as the case may be) at each step. Moreover, one can test the *rate* of convergence of the sequence or series, a crucial question that, unfortunately, is often neglected in teaching the subject in school. Let us mention just a few simple examples.

1. The sequence $\sqrt[n]{a}$, a any positive number. It is well known (Courant 1956, p. 31) that

$$\lim_{n \to \infty} \sqrt[n]{a} = 1.$$

Choose any positive number, then press the square-root key as often as you wish. It is really exciting to observe how the numbers in the display successively approach one. (We must be aware of the fact, though, that in this way we obtain the sequence $\sqrt[2^n]{a} = \sqrt[2]{a}, \sqrt[4]{a}, \sqrt[8]{a}, \ldots$, which is a partial sequence of the original one.) Moreover, when the difference between two successive terms becomes smaller than the last digit the display can hold, the final limiting value will automatically appear (in our example, the number 1). This is a vivid demonstration of the fact that, in practice, we can expect a numerical result to be of any significance only as long as the result is larger than the worst inaccuracy of the measuring device used: the final result, in our example, is displayed as 1 and not as 1.000000.

2. The geometric series $1 + 1/2 + 1/4 + 1/8 + \cdots = 2$.

Notice (table 5) that each pressing of the "+" key automatically gives the value of the partial sum that has already been accumulated (the underlined results). Thus

TABLE 5

EVALUATING $\sum_{n=0}^{k} \dfrac{1}{2^n}$

Keyboard	Display
1	1
+	1.
2	2
$1/x$	0.5
+	1.5
4	4
$1/x$	0.25
+	1.75
8	8
$1/x$	0.125
+	1.875
⋮	⋮

we can literally observe the convergence of the partial sums to their limiting value 2.

The sequence of operations can best be represented by a flow chart (fig. 1).

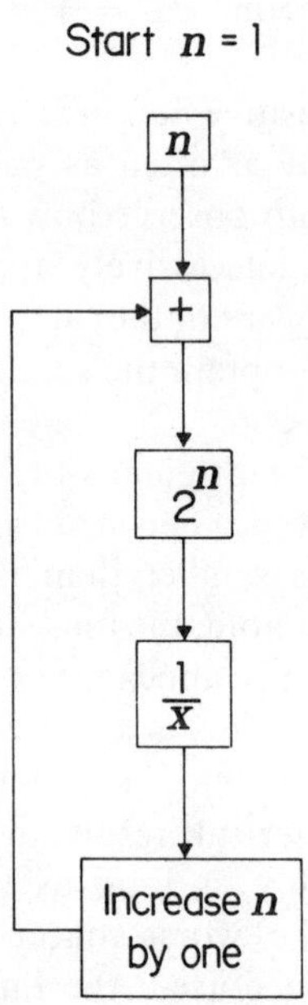

Fig. 1

It should again be noticed that the sequence of operations in figure 1 will be valid only on instruments that perform directly the addition of products; otherwise, the operations $1 + 2\ 1/x$ will be interpreted as $(1 + 2)\ 1/x$; that is, we shall get for the first two terms $1/3 = 0.3333$ instead of $1 + 1/2 = 1.5$. In such a case, we must calculate each term separately and add it to the previous terms.

We can now contrast this example with an example of a *divergent* series, such as the harmonic series $1 + 1/2 + 1/3 + 1/4 + \cdots$, and observe how the partial sums gradually increase. This would be a vivid demonstration of the fact that the condition for a series to converge, namely, that $|a_{n+1}| < |a_n|$, is necessary but not sufficient.

Other series representing important constants can likewise be tested and compared. Thus, one could compare the well-known limit for e,

$$e = \lim_{n\to\infty}\left(1 + \frac{1}{n}\right)^n,$$

with the Taylor series for e, $e = 1 + 1/1! + 1/2! + 1/3! + \cdots$, and observe how much more quickly the second expression converges than the first. Likewise, it would be interesting to compare some of the many series for π or π^2 that result from the Fourier expansion of various square and triangular functions (Courant 1956, pp. 440, 443, 446). An infinite *product*—which is much more rarely encountered than an infinite series—can also be tested; a well-known example is Wallis's product, (Courant, pp. 224, 445).

$$\frac{\pi}{2} = \frac{2}{1}\cdot\frac{2}{3}\cdot\frac{4}{3}\cdot\frac{4}{5}\cdot\frac{6}{5}\cdot\frac{6}{7}\cdots$$

As a final assignment, the class can be given the task of calculating the numerical values of some of the transcendental functions, such as $\sin x$ or $\ln x$, from their power series—in the case of $\ln x$, one must use the series for $\ln(1 - x)$ or $\ln(1 + x)$. The teacher can, at this point, tell the class how laborious and time-consuming the very same calculations must have been for the mathematicians of the previous centuries, when logarithmic and trigonometric tables were first introduced. Much historical information on this subject exists, such as the works of Klein (1953) and Smith (1953). We are actually introducing the class, through the pocket calculator, not only to numerical analysis but also to some of the more exciting chapters in the history of mathematics!

REFERENCES

Courant, Richard. *Differential and Integral Calculus*, vol. 1. London and Glasgow: Blackie & Son, 1956.

Klein, Felix. *Elementary Mathematics from an Advanced Standpoint (Arithmetic—Algebra—Analysis)*, vol. 1. New York: Dover Publications, 1953.

Smith, David Eugene. *History of Mathematics*, vol. 2. New York: Dover Publications, 1953.

USING TABLES TO TEACH MATHEMATICS

To generate a table is one thing—
the search for patterns in the numbers generated is another.

By CLIFTON KELLER

Andrews University
Berrien Springs, MI 49104

Are you looking for activities that are simple enough so that every student can contribute, even your slowest ones? The construction of tables is such an activity.

For example, you may wish to have your students construct a table listing the decimal equivalents of the reciprocals of integers. Allow your students to draw numbers from a hat or to obtain numbers by some other random process. Tell them to divide 1 by their number and continue dividing until the digits in the quotient repeat.

To help students check their work on this project and to keep them from becoming discouraged, you may wish to show them a technique involving a pocket calculator (table 1). Steps 1 and 2 obtain a partial quotient without round-off error. Steps 3 through 5 obtain the next block of numbers for the new dividend.

TABLE 1

Step 1.	Divide 1 by 17	0.058823529
Step 2.	Drop the last two digits	0.0588235
Step 3.	Multiply by 17	0.9999995
Step 4.	Subtract this from 1	0.0000005
Step 5.	Multiply by 10^7	5.0
Step 6.	Divide by 17 again	0.294117647
Step 7.	Go to step 2 (Continue until a repeating sequence is obtained.)	

For the calculations in table 1, a ten-digit TI3500 was used. Other calculators may yield slightly different numbers because of their display size and round-off rules. As a result of this exercise, students will realize that at most, $n - 1$ divisions are needed.

After the discovery of several patterns in our tables of reciprocals, one student noticed that

$$\frac{1}{98} = 0.0102040816$$

seemed to consist of ascending powers of 2. Immediately, another student saw that

$$\frac{1}{97} = 0.01030927$$

seemed to be made up of ascending powers of 3. After further investigation, we became discouraged because the patterns appeared to break down. But we were elated when we discovered that the decimal equivalent of the reciprocal of any integer from 1 to 100 could be determined by adding powers of the differences between the integer and 100. Thus 1/98 is the sum of the powers of 2 divided by the appropriate power of 10.

Example: 1/98 = 0.01000 ··· + 0.00020 ··· + 0.000004 ··· + ···

0.01
0.0002
0.000004
0.00000008
0.0000000016
0.000000000032
0.00000000000064
0.0000000000000128
0.000000000000000256
0.00000000000000000512
0.0000000000000000001024
0...

0.010204081632653 ....

Encouraged by this success, we developed the formula

$$\frac{1}{10^2 - n} = \sum_{k=0}^{\infty} \frac{n^k}{10^{2(k+1)}} = \frac{2^0}{10^{2(0+1)}} + \frac{2^1}{10^{2(1+1)}} + \frac{2^2}{10^{2(2+1)}} + \cdots$$

where n is the difference between the integer and 100. The formula worked!

In an attempt to extend the formula we noticed that

$$1/980 = 0.001020408\ldots$$

and

$$1/998 = 0.001002004\ldots$$

We could see that for numbers between 101 and 1000, we would have to modify our formula to

$$\frac{1}{10^3 - n} = \sum_{k=0}^{\infty} \frac{n^k}{10^{3(k+1)}}.$$

What about forming reciprocals in bases other than base 10? In base 2, we wrote the reciprocal of 11_2 and its logical extension.

$$\frac{1}{11_2} = \frac{1}{10_2^2 - 1} = \sum_{k=0}^{\infty} \frac{n^k}{10_2^{2(k+1)}}$$

$$= \frac{1}{100_2} + \frac{1}{10\,000_2} + \frac{1}{1\,000\,000_2} + \cdots$$

It seemed to work. We experimented with other numbers and bases. They worked! Encouraged, we wrote the formula

$$\frac{1}{10_b^q - n} = \sum_{k=0}^{\infty} \frac{n^k}{10_b^{q(k+1)}}$$

and developed the following proof:

$$\frac{1}{10_b^q - n} = \sum_{k=0}^{\infty} \left(\frac{1}{10_b^q}\right)\left(\frac{n}{10_b^q}\right)^k$$

$$\stackrel{?}{=} \frac{1}{10_b^q}\left(\frac{1}{1 - n/10_b^q}\right) = \frac{1}{10_b^q - n}$$

I have used table construction for other purposes; the results are almost as exciting as in this example. When we study factors of numbers, I post a computer printout of integers, beginning where last year's class finished. Students have their names and factors listed if they are the first to factor any number on the list. The suggestion that some answers can be found in published sources leads many students to seldom-used mathematical tables. When this resource is exhausted, students need more incentive, so greater credit is offered for each entry. Students are as thrilled when difficult numbers fall from the unfactored list as they are over major sport victories.

As an example of another activity, students are asked to make a table of integers consisting of six columns (table 2). Most students are surprised to find all prime numbers in columns one and five.

TABLE 2
Partial Table for Finding Primes

1	2	3	4	5	6
7	8	9	10	11	12
13	14	15	16	17	18
19	20	21	22	23	24

After being introduced to computers, students are asked to construct tables of squares, cubes, and so on. Once familiar with the computer, they are challenged to extend our tables of Fibonacci numbers, Fermat numbers, reciprocals, powers, roots, and prime numbers. While working on these problems, students have discovered many interesting patterns and formulated some profound questions. For example, while attempting to make a table of Fermat numbers, some students found a 1938 magazine article stating Samuel Krieger's discovery of a set of numbers that disproved Fermat's last theorem. Did he really?

I have found table construction to be extremely valuable. Slow students find their contributions important; the better students are challenged; and all students learn that by dividing a task among themselves, the group can accomplish much more than any individual.

EFFICIENT ALGORITHMS FOR THE CALCULATOR

Algorithmic design is often considered the one justification for computers in the classroom. Now this payoff comes with calculators.

By RICHARD J. FRIEDLANDER
University of Missouri–St. Louis
St. Louis, MO 63121

There have been many recent articles on using the pocket calculator as a teaching aid (e.g., Johnsonbaugh 1976, Maor 1976). These articles have mainly concentrated on what computations can be accomplished with a calculator. For the most part, however, they do not tell the user how to accomplish them. It is the purpose of this article to demonstrate some efficient ways to use the calculator, as well as to give examples that are appropriate for classroom use.

Communicating with a Calculator

In order to use any calculator, one must first learn how to communicate with it. There are two calculator "languages"—algebraic entry and reverse Polish notation (RPN). The only technical difference between the two is the order in which the four arithmetic function keys are pressed. Calculators designed for algebraic entry place the arithmetic function *between* the numbers of two-number operations. In contrast, RPN machines place the arithmetic function *after* both members of a two-number operation are entered into the machine. For example, the two-number operation 3 + 4 = 7 is entered into an algebraic calculator precisely as it appears. Pressing the keys **3**, **+**, **4**, and **=** displays the answer 7. With RPN, the same problem is solved by pressing **3**, then an **ENTER** key (also often designated as ↑), pressing **4**, and then hitting the **+** key. Since algebraic entry takes the same format as the arithmetic learned in school, almost all calculators found in classrooms today use this form of entry. Hence, in all of what follows, it will be assumed that the calculator being used is designed for algebraic entry. This assumption will be necessary in order to write down the keystrokes for the algorithms that will be presented. Appropriate modifications can be made for machines with RPN.

It will also be assumed that the calculator being used has either an automatic constant or a constant key that can be engaged in order to store a number and its associated operation for repetitive calculations. This will be vital in the next two sections. It will further be assumed that the calculator completes arithmetic operations in the order entered (as opposed to completing multiplication and division before addition and subtraction). No parentheses keys will be assumed, nor will a memory. The only function keys assumed will be **+**, **−**, **×**, and **÷**. Most calculators the author has seen in secondary school classrooms satisfy all these conditions. All computations in the article were done using such a machine with an eight-digit display.

Before presenting the algorithms, it must be pointed out that different machines may not operate alike (even assuming they satisfy all the stated assumptions). Thus, for a given calculator, it may be necessary to modify some of the algorithms. The calculations in this article have been checked on the following machines: TI1250, Litronix 2200, Novus 835, and Canon Palmtronic 8Rs.

Reciprocals

The calculator can be used in finding the reciprocal of a number *n* by pressing the keys **1**, **÷**, **n**, and **=**. If a calculation is needed to obtain *n*, this method requires the user to perform such a calculation and then record n (either mentally or on paper) before doing the division. A more efficient way, which avoids any recording, is given by the following method. Once the number n is found, enter it. Next press the ÷ key and then the = key. The number 1 should appear on the display. Then press the = key again. The desired reciprocal now appears. In short, given the number n, pressing the keys **n**, ÷, =, and = displays the reciprocal of n. (On some machines, the succession **n**, ÷, = results in a display of the reciprocal of n. Also, for machines having a constant key **k**, it is necessary to engage this key immediately after pressing the ÷ key.) For example, the problem $1 \div (36 \times 1.402 + 7.236)$ can be done wholly on the calculator without any recording when this method is used.

Powers

To find m^n (m any number, n a positive integer), one can enter m, press the × key, and then press the = key $n - 1$ times. In short, pressing the keys

$$\mathbf{m} \times \underbrace{= \cdot \cdot \cdot =}_{n-1 \text{ times}}$$

displays the calculator's answer to m^n. (For machines with a constant key **k**, the required sequence is **m**, ×, **k**, =, · · · , =.) For example, to find 1.1^{50}, enter 1.1, press the × key once and then the = key forty-nine times. The resulting display reads 117.39077. Of course, this is only an estimate for the true eight-digit, rounded-off value of 1.1^{50}, because the calculator truncates answers to eight digits at each step in the calculation. A good exercise for students would be to figure out the first n for which the calculator's displayed value of 1.1^n differs from the actual eight-digit rounded-off value of 1.1^n.

Although the method above certainly gives a good approximation for 1.1^{50}, it is not, as it stands, a very efficient way to attack the problem. However, the method can be refined by noticing that $1.1^{50} = (1.1^5)^{10}$. Entering 1.1 and pressing the × key once, followed by the = key four times, gives a value for 1.1^5. Now press the × key again, then the = key nine times. This gives a value for $(1.1^5)^{10}$. In short, pressing the keys

1.1 × = = = = × = = = = = = = = =

displays 117.39083, which is the calculator's answer to $(1.1^5)^{10}$. This answer differs in the last two digits from the one obtained for 1.1^{50}. This is to be expected, since the steps in the two calculations 1.1^{50} and $(1.1^5)^{10}$ are different. Hence different answers are being truncated along the way in the two methods.

Of course, the number of steps can be reduced by using the prime factorization of the exponent. For example, since $210 = 2 \cdot 3 \cdot 5 \cdot 7$, an approximate value of 1.0734^{210} can be found by pressing the keys

1.0734 × = × = = ×

↓ displays 1.0734^2 ↓ displays $1.0734^{2 \cdot 3}$

= = = = × = = = = = =

↓ displays $1.0734^{2 \cdot 3 \cdot 5}$ ↓ displays $1.0734^{2 \cdot 3 \cdot 5 \cdot 7}$

The final display reads 2883568.9.

A nice application of this method is the well-known limit formula

$$\lim_{n \to \infty} \left(\frac{n+1}{n} \right)^n = e\,|,$$

where $e = 2.7182818 \ldots$. By taking large composite values for n and factoring into primes, the method for powers given above makes it possible to obtain decimal approximations for $\left(\frac{n+1}{n} \right)^n$. Students can check to see that their answers are close to e. For example, if $n = 2860$, then using the factorization $2860 = 4 \cdot 5 \cdot 11 \cdot 13$, a value of 2.7170639 can be calculated for $(2861/2860)^{2860}$. In this calculation it is necessary to press the × key a total of only four times and the = key a total of twenty-nine times.

Another classroom application of the method of powers is the following paper-folding experiment. Give each student a sheet of standard notebook paper. Tell them to fold it in half; then fold it in half again; and keep folding it in half until they no longer are able to do so (this will happen after six or seven folds). If the sheet could be folded in this manner fifty times, how high would the stack be? Have each student write down a guess (in descriptive terms such as "to my desk top"). Few, if any, will guess that the stack will reach as high as the ceiling. Students can now figure out the answer with the aid of the calculator. Have them complete table 1 (given that a stack of ten sheets of standard notebook paper measures about 1 mm).

TABLE 1
Computing the Thickness of Folded Paper

Number of Folds	Number of Sheets	mm	Thickness km
0	$1 = 2^0$	0.1 mm	0.0000001 km
1	$2 = 2^1$	0.2	0.0000002
2	$4 = 2^2$	0.4	0.0000004
3	$8 = 2^3$	0.8	0.0000008
4	$16 = 2^4$	1.6	0.0000016
.	.	.	.
.	.	.	.
.	.	.	.
25	?	?	?

Using the method of powers, they can find the number of sheets after twenty-five folds, which is 2^{25}, by pressing the keys

2 × = = = = × = = = = .

The display reads 33554432. This gives 3.3554432 km. Since there are about 3 km after twenty-five folds and since thickness doubles with each fold, there are about $(2^n \cdot 3)$ km after $(25 + n)$ folds. Thus, after forty-nine folds, there are $(2^{24} \cdot 3)$ km. Using the method of powers with the factorization $24 = 2 \cdot 2 \cdot 2 \cdot 3$, this can be found by pressing the keys **2 × = × = × = × = = × 3 =**. The answer is 50331648 km. Since the distance from earth to the sun is a little less than 150 million km, it would take only two more folds, for a total of fifty-one, to reach the sun.

This fact never ceases to amaze students. It is a striking example of how fast exponential functions increase. The author has found this example to be an excellent means of conveying this idea to students. The reader will note that in solving the problem above, the work must be carefully arranged in order to prevent a calculation overflow. Also, when solving it in metric units the conversion from mm to km involves only a movement of the decimal point. Thus the tedious process of converting from inches to miles is avoided.

Parentheses

Through clever use of parentheses, some efficient algorithms can be developed for the calculator. Two particularly nice ones will be given here. One is for the addition and subtraction of fractions. The other is for synthetic division.

Notice that

$$\frac{1}{r} + \frac{1}{s} = ((1 \div r) \cdot s + 1) \div s.$$

Hence, pressing the keys

1 ÷ r × s + 1 ÷ s

displays the number $\frac{1}{r} + \frac{1}{s}$. A similar algorithm displays $\frac{1}{r} - \frac{1}{s}$. In words, "Before adding or subtracting the next fraction, multiply first by its denominator." For example, to find a value for $\frac{1}{2} - \frac{3}{7} + \frac{4}{9}$, press the following sequence of keys:

1 ÷ 2 × 7 − 3 ÷ 7 × 9 + 4 ÷ 9 =

This results in a display of 0.5158729.

One application of this method is the determination of the sum of certain infinite series. It has been shown (Johnsonbaugh 1976) that the calculator can be used to obtain approximations of the values of the partial sums of various infinite series. Using the algorithm above, the computations can be carried out without the use of a memory. For example, an approximate value of 2.7083331 for the fourth partial sum S_4 of the series $\sum_{k=0}^{\infty} \frac{1}{k!}$ can be obtained by pressing the following keys:

1 + 1 × 2 + 1 ÷ 2 × 3 × 2 + 1 ÷ 3
÷ 2 × 4 × 3 × 2 + 1 ÷ 4 ÷ 3 ÷ 2 =

Repeated application of this algorithm results in a display of 2.7182791 for S_{10}. However, a calculation overflow occurs when 11!, the next denominator, is multiplied by this value of S_{10}. By the time S_{10} has been calculated, students should suspect that the infinite series sums to e. Of course, this is by no means a proof. Indeed, there are examples of divergent series whose partial sum displays become constant on the calculator (Staib 1976). Nevertheless, the algorithm presented here for the addition and subtraction of fractions gives students a method for obtaining, on a calculator with no memory, several partial sum estimates of such infinite series as

$$\sum \frac{1}{2^k}, \quad \sum \frac{(-1)^k}{2^k}, \quad \sum \frac{1}{k(k+1)},$$

$$\sum \frac{(-1)^k}{2k+1}, \quad \text{and} \quad \sum \frac{1}{k^2}.$$

(Each of these series is convergent, their respective sums being 2, $\frac{2}{3}$, 1, $\frac{\pi}{4} = 0.7853981\ldots$, and $\frac{\pi^2}{6} = 1.6449331\ldots$.)

Through the clever use of parentheses, an algorithm for performing synthetic division on the calculator can be developed. This will be done with the following example. Let

$$f(y) = y^3 - 15y - 4,$$

which can be written as

$$\begin{aligned} f(y) &= 1 \cdot y^3 + 0 \cdot y^2 - 15y - 4 \\ &= ((1 \cdot y + 0)y - 15)y - 4. \end{aligned}$$

Hence, to evaluate $f(c)$, it is only necessary to press the following sequence of keys:

1 × c + 0 × c − 15 × c − 4 =

This has the effect of "wiping out all the exponents." As examples, the reader can check that $f(0.162) = -6.4257484$, $f(3.01) = -21.879099$, and $f(4) = 0$. Notice that the above keying sequence can be shortened by eliminating the + and 0, as well as the initial 1 and ×. However, the longer form forces the student to fill in the terms with the zero coefficient. It also establishes the following simple algorithm for evaluating a polynomial on the calculator: "Coefficient times value plus (or minus), next coefficient times value plus (or minus), ..." For these reasons, students should avoid taking such shortcuts.

Since 4 is a root of $f(y)$, the factor theorem says that $y - 4$ is a factor of $f(y)$. Using synthetic division, the quotient polynomial is obtained as shown in figure 1, where 1, 4, and 1 are the coefficients of the quadratic polynomial $y^2 + 4y + 1$ and 0 is the remainder.

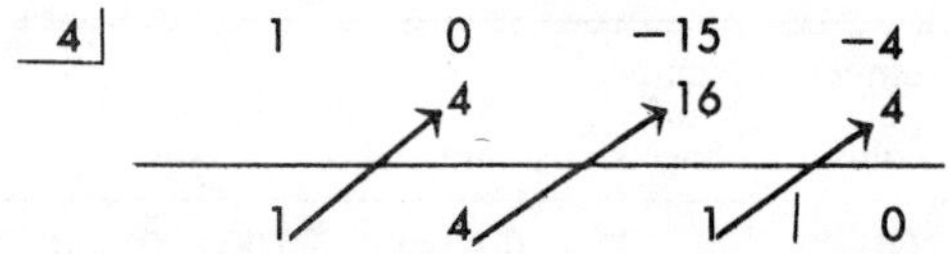

Fig. 1

Notice that in the algorithm for evaluating $f(c)$ on the calculator, synthetic division is actually being performed. For example, the quotient polynomial $(y^3 - 15y - 4) \div (y - 4)$ can be obtained by evaluating

$f(4) =$ **1 × 4 + 0 × 4 − 15 × 4 − 4 =.**

1st coefficient of quotient displayed: 1	2d coefficient of quotient displayed: 4	3d coefficient of quotient displayed: 1	Remainder displayed: 0

Thus, recording the display each time after pressing the × key will yield the coefficients of the quotient polynomial $q(y)$, where $f(y) = q(y)(y - c) + f(c)$. If $f(c) = 0$, then $q(y)$ will be a factor of $f(y)$.

Students should be given several exercises like the following:

> Use your calculator to check if the given number c is a root of the given polynomial $f(y)$. In each case, find the quotient polynomial $q(y)$ and remainder r satisfying $f(y) = q(y)(y - c) + r$ (remember to fill in all terms having zero coefficient).

c	$f(y)$
2	$2y^3 - 5y^2 + y + 2$
3	$2y^4 - 7y^3 + 5y^2 - 7y + 3$
5	$y^5 - 9y^4 + 8y^3 + 60y^2 + y - 5$
−15	$y^3 + 16y^2 + 15y + 1$
−2	$6y^8 + 13y^7 - 4y^5 + 8y + 16$

It has been my experience that many students forget synthetic division over a period of time and are forced to fall back on long division of polynomials. This is a shame, since the latter method takes so much longer. However, synthetic division on the calculator is so easy, quick, and natural that students who learn it with the aid of the machine are more likely to retain the skill in the long run.

Summary

In order for students to fully realize what can be done with the aid of the pocket calculator, they must first learn how to use the calculator in the most efficient manner. This article has presented some shortcuts that can be readily applied to classroom work at the secondary level. Although it is useful for students to master such algorithms as these, it is equally important that they understand why these methods work. Blindly learning efficient means of calculator operation will not serve them well in the long run. A deeper understanding of the reasons behind these methods, as well as the limitations of their use, must be the goal. Teachers who bring pocket calculators into their classrooms must be aware of this. Only then will the full value of the calculator as a teaching aid be realized.

REFERENCES

Johnsonbaugh, Richard. "Applications of Calculators and Computers to Limits." *Mathematics Teacher* 69 (January 1976):60–65.

Maor, Eli. "The Pocket Calculator as a Teaching Aid." *Mathematics Teacher* 69 (October 1976): 471–75.

Staib, John. "Reader Reactions." *Mathematics Teacher* 69 (October 1976):437.

THE ROLE OF PROGRAMMABLE CALCULATORS AND COMPUTERS IN MATHEMATICAL PROOFS

Data generated by computing devices may be used as an essential part of a mathematical proof.

By STEPHEN L. SNOVER
University of Hartford
West Hartford, CT 06117

and MARK A. SPIKELL
Lesley College
Cambridge, MA 02138

Computers and programmable calculators are becoming increasingly important as aids in the creation of mathematical proofs. They not only assist in discovering data from which conjectures can be made, but they also can complete a proof by exhaustively searching a (large) finite collection of cases not included by an already established argument. A striking example illustrating these applications is the recent solution of the famous four-color map conjecture. In 1976 Appel and Haken proved that four colors are sufficient to color any plane map, basing their argument on computer-obtained results that would be virtually impossible for a person to verify by hand (1977).

To consider how machines can be used in the creation of mathematical proofs, one does not need to explore advanced problems. Instead, this article illustrates the role of programmable calculators and computers in mathematical proofs by exploring a simple problem from number theory. The problem is based on a puzzle found in Dudeney (1967, p. 43) and is suitable for presentation to high school and college students.

The problem

The integer 153 has the curious property that it is equal to the sum of the cubes of its digits, that is,

$$153 = 1^3 + 5^3 + 3^3.$$

Are there other positive integers with this property? If so, is there a pattern, rule, function, or algorithm by which they may be generated?

Perhaps a wise beginning is to find any other three-digit integers with the cubed digit sum property. However, attempts to find them by hand or even with a four-function calculator are tedious and time-consuming. To check all 900 three-digit integers could require 4500 arithmetic calculations, since five computations must be completed to test each three-digit integer—each of the three digits must be cubed and two additions completed to form the sum. This is an awesome exercise for even the most dedicated number manipulation enthusiast. But programmable pocket calculators and computers can perform numerous arithmetic calculations with ease and speed.

Designing programs for discovering data

In order to write a machine program to locate all three-digit integers with the cubed digit sum property, it is helpful to express the desired number property algebraically. If n has three digits, then

$$n = 100a + 10b + c$$

where a, b, and c are the hundreds, tens, and units digits, respectively. Consequently, if n is equal to the sum of the cubes of its digits, then

$$100a + 10b + c = a^3 + b^3 + c^3.$$

With this equation it is possible to test whether any three-digit integer n equals the sum of the cubes of its digits.

There are actually many techniques for generating all three-digit integers n and their respective digits. As revealed by the flowchart in figure 1, the authors have chosen a programming style that first generates each n and then pulls off its digits one at a time. Used in the flowchart are the functions, INT( ) and FRAC( ), which take the integer part or fractional part, respectively, of the value inside the parentheses. Actual programs in BASIC computer language and in machine language for the Texas Instruments SR 56 and TI 57 and for the Hewlett-Packard HP 25 programmable calculators are presented in the appendix.

Running any of these programs produces just four three-digit numbers that equal the sums of the cubes of their respective digits. These numbers are 153, 370, 371, and 407.

Altering programs to obtain all results for four or fewer digits

It is not difficult to alter any of the programs to search for more data, because the original flowchart in figure 1 was designed with this flexibility in mind. In fact, it is only necessary to change the starting and ending values in the initialization. For example, to find all four-digit numbers with the cubed digit sum property, merely change the starting and ending values from 100 and 999 to 1000 and 9999, respectively.

Running these programs to test all four-, one-, and two-digit integers yielded only two more integers with the desired property—the single digit numbers 0 and 1.

Conjectures

These results do not reveal any obvious way to generate all integers with the cubed digit sum property. Could it be that there are no other such integers? Could it be that only integers with an odd number of digits can have the cubed digit sum property? In other words, perhaps one or the other of these conjectures is true:

CONJECTURE 1. *There exist no integers with five or more digits with the cubed digit sum property.*

CONJECTURE 2. *All integers with the cubed digit sum property have an odd number of digits.*

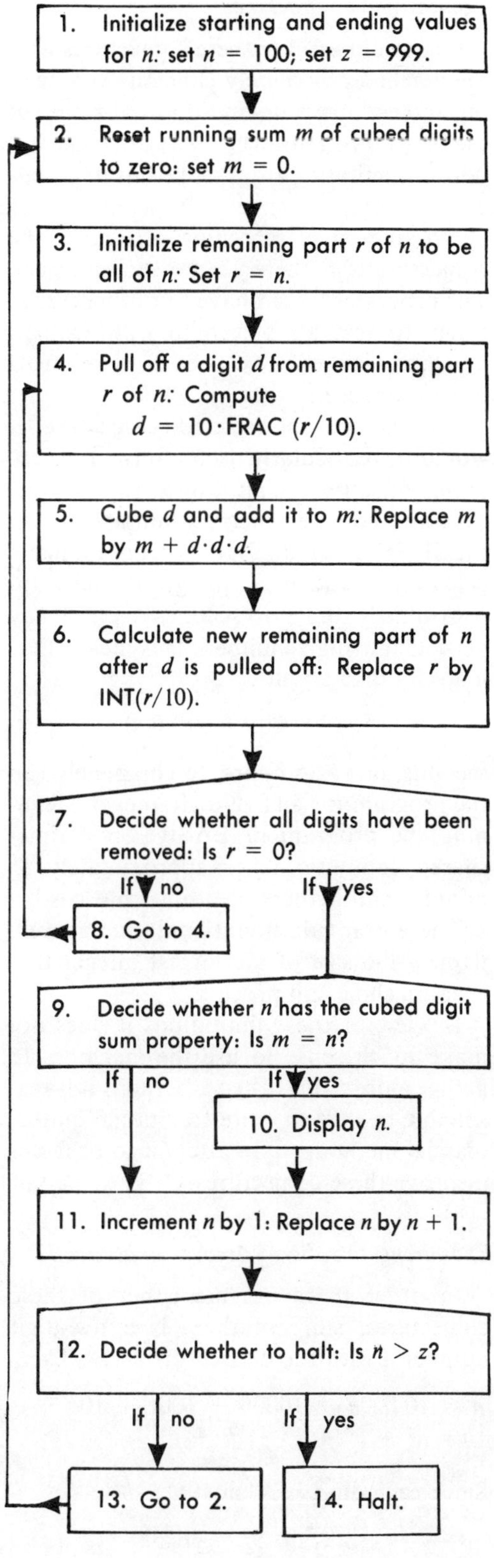

Fig. 1

Limitations of machines

One could explore these conjectures with the machines by merely changing the starting and ending values in the initialization of the programs. However, there are some important limitations that make this impractical.

No amount of searching will affirm the conjecture that there are no such integers. The programs would have to run forever in order to test all possibilities. Machines, however, can only be used to examine a finite number of cases.

To explore the second conjecture it would be particularly useful to run the programs for all five-digit numbers. However, time and cost become real limitations. On the SR 56, TI 57, and HP 25 programmable calculators, the program given takes approximately $(0.7 + k)$ seconds to check any particular k-digit number. This means that it would take about

$$5.7(90\ 000) = 513\ 000$$

seconds, or 142.5 hours, to completely run the program for all five-digit integers. Running the program in BASIC on a time-shared computer takes approximately 15 minutes and is therefore somewhat costly.

There is an additional limitation as well, namely the size of the largest integer that each machine can process.

In view of these limitations it does not make much sense to use the machines to gather more data. Instead, perhaps it is reasonable to shift to more traditional mathematical methods in an attempt to prove or disprove these conjectures.

Examining five-digit numbers analytically

In order to investigate either of these conjectures, suppose there is a five-digit number n with the desired property. Then

$$\begin{aligned} n &= 10\ 000a + 1000b + 100c + 10d + e \\ &= a^3 + b^3 + c^3 + d^3 + e^3. \end{aligned}$$

Since each digit is at most 9,

$$n \leq 5 \cdot 9^3 = 3645.$$

But $n \leq 3645$ contradicts the fact that n has five digits. Hence, there is no five-digit number with the cubed digit sum property.

Does this indicate that conjecture 1 might be true? If so, perhaps similar reasoning will show that no number of five or more digits can equal its cubed digit sum.

An analytical proof of conjecture 1

Assume n has the desired property and suppose n has k digits where $k \geq 5$. Then

$$k = \text{INT}(\log_{10}(n)) + 1,$$

where INT( ) is the function that takes the integer part of the value inside the parentheses. This expression for k is then greater than or equal to 5, because $n \geq 10^5$. Since each digit is at most 9,

$$n \leq k \cdot 9^3 < k \cdot 10^3.$$

By analyzing the function $\log_{10}(x)$, it can be shown that

$$\log_{10}(x) + 4 < x$$

for all $x > 4.67$ (see fig. 2). Since $k \geq 5$, it follows that

$$\log_{10}(k) + 4 < k.$$

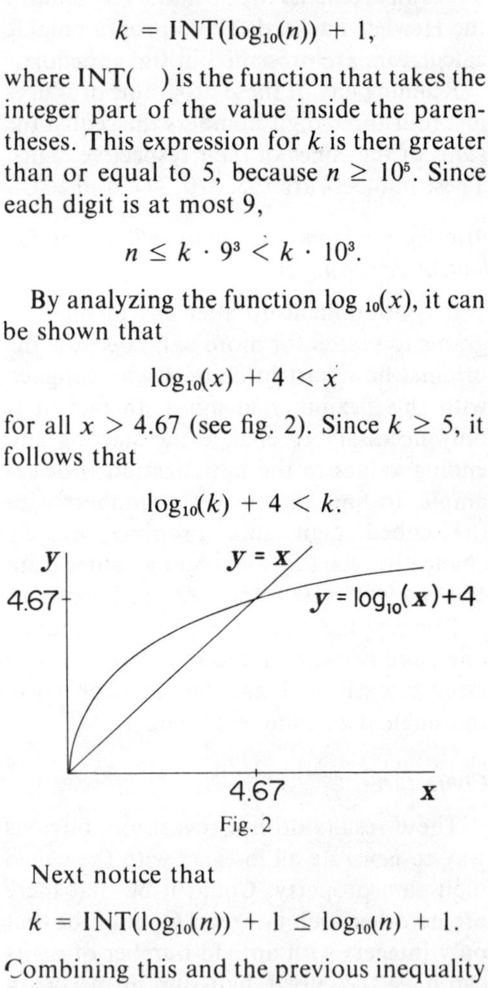

Fig. 2

Next notice that

$$k = \text{INT}(\log_{10}(n)) + 1 \leq \log_{10}(n) + 1.$$

Combining this and the previous inequality produces

$$\log_{10}(k) + 4 < \log_{10}(n) + 1.$$

Then it follows that

$$\log_{10}(k) + 3 < \log_{10}(n),$$

$$\log_{10}(k \cdot 10^3) < \log_{10}(n),$$

and finally,

Therefore,

$$n \leq k \cdot 9^3 < k \cdot 10^3 < n,$$

which is a contradiction. This forces the conclusion that no k-digit number, where $k \geq 5$, has the desired property. Conjecture 1 is now proved.

Limitations for the analytical proof

Although the proof for conjecture 1 is valid when n has five or more digits, the same line of reasoning fails when n has four or fewer digits. Specifically, if n has four digits,

$$n \leq 4 \cdot 9^3 = 2916;$$

and since 2916 is a four-digit number, no contradiction is reached.

The complete proof

Importantly, the analytical proof for five or more digits (conjecture 1), together with the results obtained for four-digit or smaller numbers using the machines, constitutes a complete mathematical proof that there exist exactly six nonnegative integers with the cubed digit sum property, namely, 0, 1, 153, 370, 371, and 407. Neither the analytical part nor the machine-search part of the proof forms a proof on its own. (It is interesting to note that the same proof verifies conjecture 2, which is true as stated.)

Conclusion

Programmable calculators and computers should be considered important mathematical tools that are useful in creating mathematical proofs. Since these machines can examine numerous possibilities in a relatively short period of time, they are powerful data-gathering instruments. Moreover, the data generated by these machines may be used as an essential part of a mathematical proof, as this article illustrates.

Until recently most mathematicians and mathematics educators have resisted using machine-produced results as legitimate parts of mathematical proofs. As a result of the work done on the four-color map problem, a new type of mathematical proof should be more widely accepted. People should be encouraged to look for other problems that can be proved by some combination of calculator/computer calculation and mathematical analysis. Here for interested readers is a small collection of such problems, all simply stated and from number theory.

Additional problems

1. Find all integers that equal the sums of the squares of their digits.

2. Find all integers that equal the sums of the fourth powers of their digits.

3. Notice that

$$1233 = 12^2 + 33^2$$

but

$$12^2 + 56^2 + 33^2 \neq 125633.$$

Find all integers with an even number of digits and with the property that when broken up two digits at a time, the sum of all the squares is the original integer.

4. Notice that $145 = 1! + 4! + 5!$. Find all such integers.

5. Notice that $36 = 3 \cdot 2 + 6 \cdot 5$. Find all integers that equal the sum of the products of each digit with one less than that digit, that is

$$n = \sum d \cdot (d - 1)$$

where the summation is over all d of n.

6. The number 171 equals the sum of the squares of the first, seventh, and first Fibonacci numbers, that is,

$$171 = f^2(1) + f^2(7) + f^2(1)$$

$$= 1^2 + 13^2 + 1^2$$

where the Fibonacci numbers are given by $f(0) = 0$, $f(1) = 1$, and

$$f(m) = f(m - 1) + f(m - 2)$$

for all $m \geq 2$. Find all such numbers.

7. In base 8, $134_8 = 1^3 + 3^3 + 4^3$. Find all base 8 integers with the cubed digit sum property.

Epilogue

In this article it is proved that there are only six (base 10) integers with the cubed digit sum property. The analytical part of the proof is valid only for integers with five or more digits. Interested readers may find it intriguing to try to find an analytical proof that covers all cases and does not need to refer to a machine search in order to obtain a complete proof.

APPENDIX

Programs That Follow the Flowchart in Figure 1.

Program for BASIC

```
LIST
CUBES

10  DATA 100,999
20  READ N,Z
30  LET M=0
40  LET R=N
50  LET D=10*(R/10-INT(R/10))
60  REM ************************
70  REM SUBTRACTION GIVES ROUND-
80  REM OFF ERRORS, THUS STEP 100
90  REM ************************
100  LET D=INT(D+.5)
110  LET M=M+D*D*D
120  LET R=INT(R/10)
130  IF R <> 0 THEN 50
140  IF N <> M THEN 160
150  PRINT N
160  LET N=N+1
170  IF N <= Z THEN 30
180  END
```

Program for the Texas Instruments SR 56

step of flowchart	*program*	
1.	See initialization sequence below.	
2.	00 0 01 STO 02 2	03 $x \rightleftarrows t$
3.	04 RCL 05 1 06 2nd Pause	07 STO 08 3
4.	09 RCL 10 3 11 ÷ 12 1 13 0 14 = 15 STO	16 4 17 INV 18 2nd Int 19 × 20 1 21 0 22 =
5.	23 × 24 x^2 25 =	26 SUM 27 2
6.	28 RCL 29 4 30 2nd Int	31 STO 32 3
7&8.	33 INV 34 2nd $x = t$	35 0 36 9
9.	37 RCL 38 1 39 $x \rightleftarrows t$ 40 RCL 41 2	42 INV 43 2nd $x = t$ 44 4 45 7
10.	46 R/S	
11.	47 1 48 SUM 49 1	50 RCL 51 1 52 $x \rightleftarrows t$
12&13.	53 RCL 54 0 55 2nd $x \geq t$	56 0 57 0
14.	58 0 59 R/S	

Memory usage: $R_0 = z =$ end value, $R_1 = n$, $R_2 = m =$ cube sum, $R_3 = r$, $R_4 = r/10$.

Initialization: 2nd CMs, RST, 100, STO 1, 999, STO 0, R/S.

Program for the Texas Instruments TI 57 and Radio Shack EC 4000

step of flowchart	*program*	
1.	See initialization sequence below.	
2.	00 2nd Lbl 1 01 0 02 STO 2	03 STO 7
3.	04 RCL 1 05 2nd Pause 06 STO 3	
4.	07 2nd Lbl 2 08 RCL 3 09 ÷ 10 1 11 0 12 =	13 STO 4 14 INV 2nd Int 15 × 16 1 17 0 18 =
5.	19 × 20 x^2	21 = 22 SUM 2
6.	23 RCL 4 24 2nd Int 25 STO 3	
7&8.	26 INV 2nd $x = t$ 27 GTO 2	
9.	28 RCL 1 29 STO 7 30 RCL 2	31 INV 2nd $x = t$ 32 GTO 3
10.	33 R/S	
11.	34 2nd Lbl 3 35 1	36 SUM 1 37 RCL 1
12&13.	38 STO 7 39 RCL 0	40 2nd $x \geq t$ 41 GTO 1
14.	42 0 43 R/S	

Memory usage: $R_0 = z =$ end value, $R_1 = n$, $R_2 = m =$ cube sum, $R_3 = r$, $R_4 = r/10$, $R_7 =$ test register.

Initialization: INV 2nd C.t, RST, 100, STO 1, 999, STO 0, R/S.

Program for the Hewlett-Packard HP 25 and HP 25C

step of flowchart	*program*	
1.	See initialization sequence below.	
2.	00 (R/S) 01 0 02 STO 2	
3.	03 RCL 1 04 STO 3	
4.	05 RCL 3 06 1 07 0 08 ÷ 09 STO 4	10 g FRAC 11 1 12 0 13 ×
5.	14 ENTER 15 g x^2	16 × 17 STO + 2
6.	18 RCL 4 19 f INT	20 STO 3
7&8.	21 g $x \neq 0$ 22 GTO 05	
9.	23 RCL 1 24 RCL 2	25 f $x \neq y$ 26 GTO 28
10.	27 R/S	
11.	28 1 29 STO + 1	
12&13.	30 RCL 1 31 RCL 0	32 f $x \geq y$ 33 GTO 01
14.	34 0 35 GTO 00	

Memory usage: $R_0 = z =$ end value, $R_1 = n$, $R_2 = m =$ cube sum, $R_3 = r$, $R_4 = r/10$.

Initialization: f PRGM, f REG, 100, STO 1, 999, STO 0, R/S.

BIBLIOGRAPHY

Appel, Kenneth, and Wolfgang Haken. "The Solution of the Four Color Map Problem." *Scientific American,* October 1977, pp. 108–21.

———. "Every Planar Map Is Four Colorable: Part 1, Discharging." *Illinois Journal of Mathematics* 21 (September 1977): 429–90.

———. "Every Planar Map Is Four Colorable: Part 2, Reducibility." *Illinois Journal of Mathematics* 21 (September 1977): 491–567.

Dudeney, Henrey Earnest. *536 Puzzles & Curious Problems.* Edited by Martin Gardner. New York: Charles Scribner's Sons, 1967.

Saaty, Thomas, and Paul Kainen. *The Four Color Problem: Assault and Conquest.* New York: McGraw-Hill, 1977.

FINDING AREAS UNDER CURVES WITH HAND-HELD CALCULATORS

No, this is not the usual limit of the sum from calculus. All you need are some points on the curve—and, of course, your calculator.

By ARTHUR A. HIATT

California State University–Fresno
Fresno, CA 93740

While the debates about the hand-held calculator go on, perhaps its real significance is being overlooked. Clearly, we now have a tool that can assist mathematics educators to focus on one of the most neglected areas of mathematics instruction, that is, the method of inquiry used in mathematics. We tend to overemphasize content, whereas an equally important aspect of mathematics is its method of inquiry.

As a minimum, mathematics instruction should help the student develop the ability—

1. to make observations
2. to organize observations (data)
   a) to recognize patterns
   b) to make conjectures
3. to specialize and generalize
   a) to use inductive reasoning
   b) to reason by analogy
4. to invent symbolism to express mathematical ideas
   a) to accept conventional symbolism
5. to prove conjectures
   a) to invent or accept an axiomatic structure.

It is the intent of this paper to apply the method of inquiry above to a problem in secondary school geometry—the area of a circle. Often the approach is quite intuitive, showing that the area of a regular inscribed polygon approaches the area of a circle as the number of sides of the polygon increases (Jacobs & Myer 1972, p. 595). Since the calculation of the area of a regular polygon becomes increasingly tedious as the number of sides increases and requires trigonometry, most teachers rely on the students' intuitive idea of limits. In general, the student has to accept the fact that the circumference of a circle is given by $2\pi r$.

A more primitive, but perhaps more instructive, method is to trace circles of various radii on graph paper and use a simple balance to weigh them. I have used this method with secondary general mathematics students to determine the constant in $A = kr^2$ (formula for the area of a circle). The largest average value of k was 3.34. This value was obtained with a balance made from a straw and three pins. One pin is used as a fulcrum, one holds the cut-out circle, and one holds the comparable amount of squared paper (figure 1). A value of k using a simple commercial balance was found to be close to 3.1. This same method was used by Galileo in about 1600 to predict the formula for the area of a cycloid (Eves 1969, p. 295). Galileo conjectured the area of a cycloid to be about three times the area of the generating circle. The first published proof that the area is exactly three times the generating circle was in 1644 by Evangelista Torricelli, a student of Galileo. The method of proof involved the use of infinitesimals (the early beginnings of integral calculus).

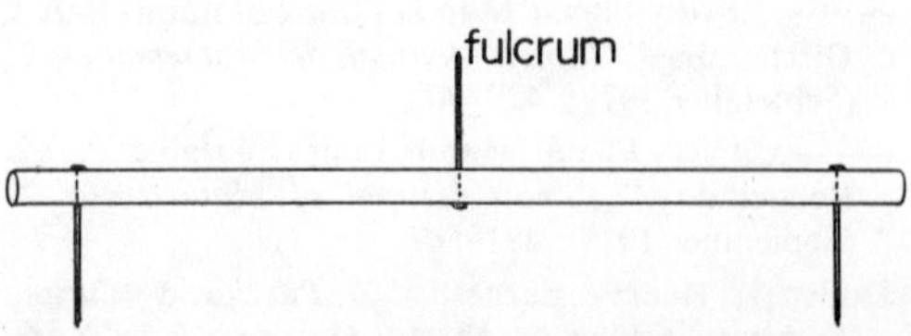

Fig. 1

It is informative to recapitulate the method of Galileo, for it emphasizes the method of inquiry used in mathematics. First he made several observations about the area of cycloids (weighings of various cycloids generated by circles of various radii). After carefully organizing his data, he conjectured that the area of a cycloid is approximately three times the area of the generating circle ($A_c \approx 3\pi r^2$). Finally, new axioms and definitions had to be developed to construct a formal proof. His student, Torricelli, gave us this proof. Thus we see how content ($A_c = 3\pi r^2$) is added to the body of mathematical knowledge. Clearly, the method of discovery is in many cases as exciting as the actual fact (content), if not more so.

Let us return to the area of a circle. The appendix indicates how to calculate the area of polygons, given the coordinates of the vertices (Hiatt 1972, p. 598). The area of a polygon is given by

$$(1)\quad A = \frac{1}{2}\left[\sum_{i=1}^{n-1}(x_i y_{i+1} - y_i x_{i+1}) + x_n y_1 - y_n x_1\right]$$

where starting with any vertex (x_1, y_1) the other vertices are numbered (x_2, y_2), etc., in a counterclockwise manner. As an example, consider figure 2. What is the area of *ABCD*?

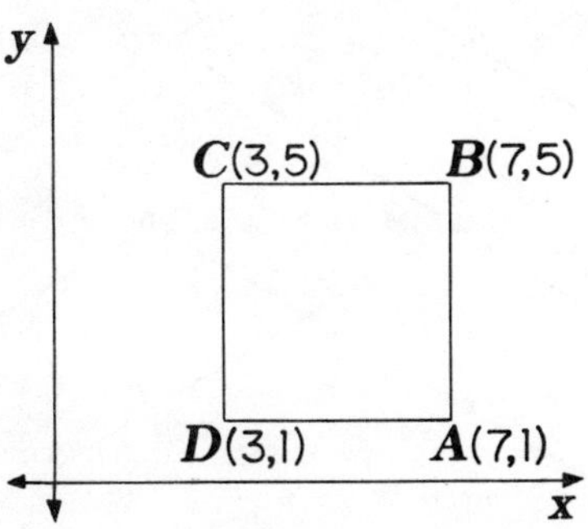

Fig. 2

Clearly the area is $4 \times 4 = 16$ square units.

By equation (1),

$$A = \frac{1}{2}\left[7\cdot 5 + 7\cdot 5 + 3\cdot 1 + 3\cdot 1 - 1\cdot 7 - 5\cdot 3 - 5\cdot 3 - 1\cdot 7\right]$$

$$= \frac{1}{2}(76 - 44) = 16.$$

We now have a powerful method to determine the area under any curve. Suppose we want the area under the curve as in figure 3.

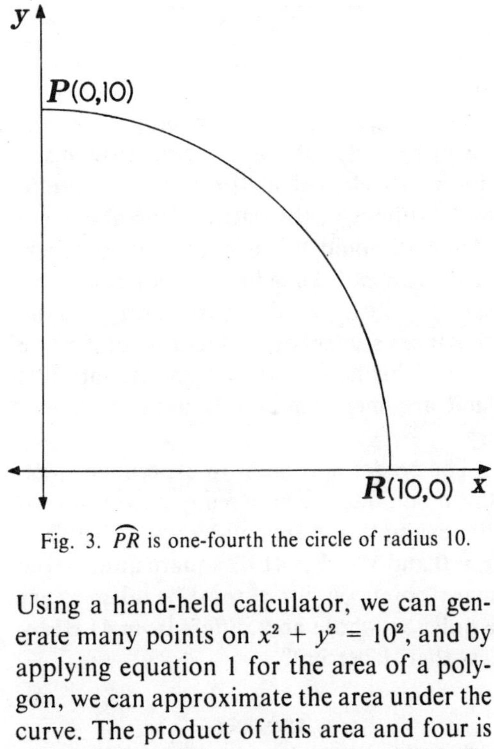

Fig. 3. $\overparen{PR}$ is one-fourth the circle of radius 10.

Using a hand-held calculator, we can generate many points on $x^2 + y^2 = 10^2$, and by applying equation 1 for the area of a polygon, we can approximate the area under the curve. The product of this area and four is approximately 100π square units. With the points (0, 0), (10, 0), (9, 4.36), (8, 6), (7, 7.14), (6, 8), (5, 8.66), (4, 9.16), (3, 9.54), (2, 9.80), (1, 9.95), and (0, 10), we get 310 square units for an approximation to the area of the circle. This is equivalent to finding the area of a forty-sided polygon inscribed in the circle. But with a hand-held calculator it is easy to get many ordered pairs, including the following, (9.5, 3.12), (8.5, 5.27), (7.5, 6.61), (6.5, 7.60), (5.5, 8.35), (4.5, 8.93), (3.5, 9.37), (2.5, 9.68), (1.5, 9.89), and (0.5, 9.99). If a polygon is drawn that connects all twenty-two of the preceding coordinates, its area is 78.2 square units. Therefore, the approximate area of the circle is 313 square units.

In a class with more than one calculator available, students could complete table 1.

TABLE 1

Radius (r)	Area of approximating polygon		Area ÷ r^2
10	(40 sides)	310	3.10
10	(80 sides)	313	3.13
	⋮		⋮

More enterprising classes may want to investigate the average length of segments from the origin to the perpendicular bisector of the sides of the polygon. Obviously, the hand-held calculator is a very useful tool; without it, the data for the above experiment would be very difficult to generate, because of the tedious calculations necessary. Such a demonstration usually convinces students that the area of a circle is πr^2. This helps them accept the intuitive limit argument commonly used in geometry.

The reader may want to experiment with this method to see how really powerful it is. For example, the area under $y = x^2$ between $x = 0$ and $x = 5$ is 41.67 square units. How many points on $y = x^2$ must be taken to get a polygon whose area differs from 41.67 by less than 1 percent?

APPENDIX

Areas from the Coordinates of the Vertices

We develop a formula for the area of a triangle when the coordinates of its vertices are given in a coordinate plane. Since any convex polygon can be divided into a finite number of triangles, the method can be generalized to include all convex polygons.

Let triangle ABC be given as below and compute the areas as shown:

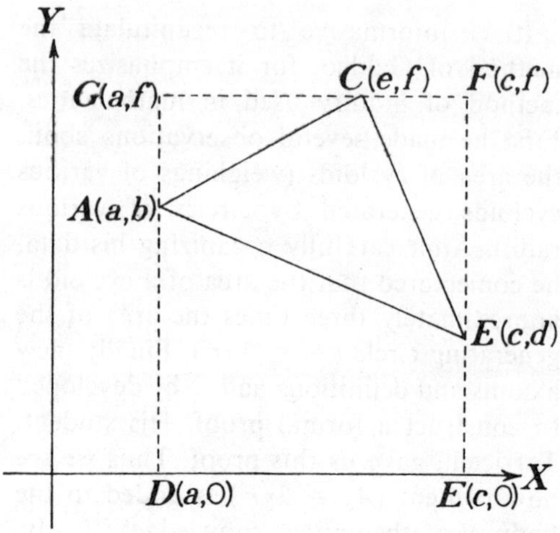

$$\begin{aligned}\text{area } \Delta ABC &= \text{area } ABFG \\ &\quad - \text{area } \Delta GAC - \text{area } \Delta CBF \\ &= \tfrac{1}{2}(AG + BF)(GC + CF) \\ &\quad - \tfrac{1}{2}\cdot AG\cdot GC - \tfrac{1}{2}\cdot CF\cdot BF \\ &= \tfrac{1}{2}(AG\cdot CF + BF\cdot GC) \\ &= \tfrac{1}{2}[(f-b)(c-e) + (f-d)(e-a)] \\ &= \tfrac{1}{2}(cf + eb + ad - de - af - bc).\end{aligned}$$

Notice the pattern: three positives and three negatives. If we list the ordered pairs of the vertices in the following manner

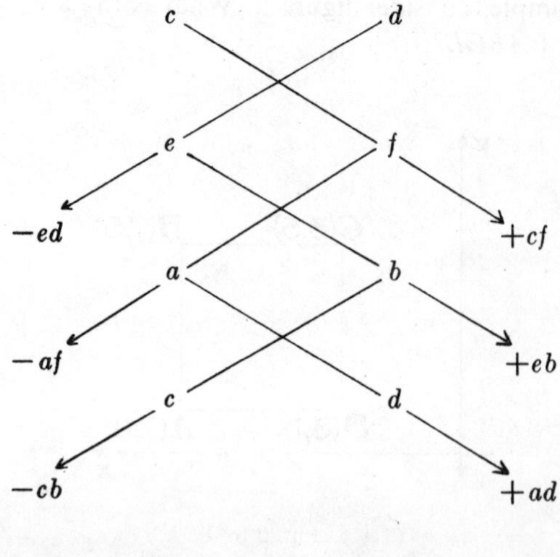

and form the products as indicated, we get a formula for the area of ΔABC:

$$\text{area } \Delta ABC = \frac{1}{2}\begin{vmatrix} c & d \\ e & f \\ a & b \\ c & d \end{vmatrix}$$

$$= \tfrac{1}{2}(cf + eb + ad - ed - af - cb).$$

The rule is simple. Start at any ordered pair. Go in a counterclockwise direction and affix the first ordered pair to the end of the array. In our example, (c, d) appears at the top and at the bottom. If we had an n-gon, we would have $(n + 1)$ ordered pairs in our array; the procedure for multiplying remains the same. The reader may enjoy proving the formula for the n gon.

REFERENCES

Eves, Howard. *History of Mathematics*. New York: Holt, Rinehart & Winston, 1969.

Hiatt, Arthur. "Problem Solving in Geometry." *Mathematics Teacher* 65 (November 1972):595–600.

Jacobs, Russell, and Richard Meyer. *Discovering Geometry*. New York: Harcourt Brace Jovanovich, 1972.

APPLICATIONS OF CALCULATORS AND COMPUTERS TO LIMITS

Here is a concise treatment of machine-assisted limit theory complete with relevant inequalities and a discussion of two "frustrating" series.

By Richard Johnsonbaugh
Chicago State University
Chicago, Illinois

A COMPUTER or calculator is sometimes regarded as a supplementary tool that affords an alternative and easier method of computation. For example, a student may use a computer to find numerical results to a problem without improving his theoretical grasp of the problem. There is, however, another point of view. If a student looks at how the computer interacts with the problem he may gain insight into the theoretical aspects of the problem. In this article I will give some examples from the elementary theory of limits to show how a computer or calculator can be used as a motivating, essential device, integrated with the theory.

Representation of Numbers by a Machine

This brief section will give the assumptions concerning the representation of a number by a machine. This discussion provides the basis for the discussion of the sum of a series.

A machine can display only a fixed number of digits, which will be consistently denoted in this paper by m. For many calculators this setting is $m = 8$ digits and for many computers this setting is $m = 6$ digits. For the purposes of this exposition, it will be assumed that the machine simply drops all digits past the first m. For example, if $m = 3$, the numbers 0.333, 0.33318, 0.3337, and $1/3 = 0.333\ldots$ will all appear as 0.333. Many calculators now available have keys to select the desired value of m.

Still assuming $m = 3$, suppose a number L is represented as $M = 0.d_1d_2d_3$. How large an error can occur between what we see, M, and the actual number, L? That is, what is the largest possible value that $|L - M|$ can assume? The number L lies between $0.d_1d_2d_3000\ldots$ and $0.d_1d_2d_3999\ldots$. Thus the maximum value that $|L - M|$ can assume is

$$\begin{aligned} 0.d_1d_2d_3999\ldots - 0.d_1d_2d_3000\ldots \\ = 0.000999\ldots = 0.001 \\ = 1/10^3 = 1/10^m. \end{aligned}$$

By the same reasoning we can conclude that this formula holds for any value of m. Thus, if a number L is represented as $M = 0.d_1d_2d_3 \ldots d_m$, then $|L - M| \leq 1/10^m$. If it is assumed that the machine rounds off numbers, the same conclusion, $|L - M| \leq 1/10^m$, results; although of course the argument is changed somewhat.

The Sum of an Infinite Series

Given a sequence $\{a_k\}$, we form the sequence of partial sums $\{s_k\}$ given by $s_k = a_1 + a_2 + \ldots + a_k$. Suppose a computer or calculator is programmed in such a way that each command to the machine sums another term of the series. For example, if

I should like to thank the referees and Professor Michael Sullivan of Chicago State University for their very helpful comments and suggestions concerning this article.

$a_k = 1/3^k$, the output appears as in table 1 for $m = 3$ and $m = 6$. Taking $a_k = (-1)^k$ and $m = 3$, the output appears as in table 2.

After experimenting with several examples like those above, the student should be led to think about what is meant by the sum of an infinite series. Examples should include the geometric series with which the student has probably already had some experience, and several divergent series. Each series should be studied for various values of m. The conclusion to be drawn is that a sum for a series is a value that has the property that eventually it does not change even though additional terms are added. In table 1, the value does not change after six

TABLE 1

Values of the Partial Sum $s_k = \sum_{i=1}^{k} \frac{1}{3^i}$

	$m = 3$	$m = 6$
s_1	0.333	0.333333
s_2	0.444	0.444444
s_3	0.481	0.481481
s_4	0.493	0.493827
s_5	0.497	0.497942
s_6	0.499	0.499314
s_7	0.499	0.499771
s_8	0.499	0.499923
s_9	0.499	0.499974
s_{10}	0.499	0.499991
s_{11}	0.499	0.499997
s_{12}	0.499	0.499999
s_{13}	0.499	0.499999
s_{14}	0.499	0.499999
.	.	.
.	.	.
.	.	.

partial sums for $m = 3$ and does not change after twelve partial sums for $m = 6$. In table 2, it is apparent that the value always changes so that this series does not have a sum.

TABLE 2

Values of the Partial Sum $s_k = \sum_{i=1}^{k} (-1)^i$

	$m = 3$
s_1	−1.000
s_2	0.000
s_3	−1.000
s_4	0.000
.	.
.	.
.	.

The sum of a series should be independent of the machine and the size of the display. Thus the sum of a series is a number L that has the property that for any size display, eventually the sum is displayed identically with L. In the earlier example, the sum of the series

$$\sum_{k=1}^{\infty} 1/3^k$$

is $0.4999\ldots = 0.5 = 1/2$.

The relevant inequalities can now be introduced if desired. If L is the sum of a series for a display m, then the sum L and a partial sum s_k eventually appear identically. Thus eventually we have $|s_k - L| \leq 1/10^m$. The student is led to conclude that "eventually" means that $k \geq K$ for some integer K. In table 1, we may take $K = 6$ (or larger) for $m = 3$ and $K = 12$ (or larger) for $m = 6$. Therefore, if L is the sum of the series Σa_k, then for any m, there exists K such that $|s_k - L| \leq 1/10^m$ for $k \geq K$. Notice that in the traditional definition "for every $\epsilon > 0$" is replaced by "for any size display" or "for any m," and "$k \geq K$" is replaced by "eventually."

A Convergence Test

Once students understand the idea in the previous section, they are ready for other results, some normally studied in advanced calculus. The convergence test that follows is developed along exactly the same lines as in the previous section.

The test is that a series, Σa_k, where a_k is nonnegative for every k, converges if the partial sums $s_k = a_1 + \ldots + a_k$ are bounded. To illustrate the argument, consider the series

$$\sum_{k=1}^{\infty} 1/k!.$$

Since for $k \geq 2$,

$$2^{k-1} = 2\cdot 2 \cdots 2\cdot 2$$
$$\leq k\cdot(k-1)\cdots 3\cdot 2 = k!,$$

it follows that $1/k! \leq 1/2^{k-1}$ and thus

$$s_k = \frac{1}{1!} + \frac{1}{2!} + \frac{1}{3!} + \cdots + \frac{1}{k!} \leq$$

$$1 + \frac{1}{2} + \frac{1}{2^2} + \cdots + \frac{1}{2^{k-1}}$$

$$= 2\left(1 - \frac{1}{2^k}\right) \leq 2.$$

Thus the partial sums of $\Sigma 1/k!$ are bounded by 2. The first partial sum s_1 is 1.000 (see table 3). As additional terms are added, the partial sums increase, since the terms of the series are positive. Suppose the ones' place changes infinitely often as additional terms are added. Then, since it changes at least two times and since each change increases the sum by at least 1, after two changes the display would be at least 3.000.

TABLE 3

Values of the Partial Sum $s_k = \sum_{i=1}^{k} \frac{1}{i!}$

	$m = 3$
s_1	1.000
s_2	1.500
s_3	1.666
s_4	1.708
s_5	1.716
s_6	1.718
s_7	1.718
s_8	1.718
.	.
.	.
.	.

But this is impossible, since the partial sums are bounded by 2. Therefore, eventually the ones' place remains constant. Actually, the ones' place never changes at all. Next, consider the tenths' place. Suppose the tenths' place changes infinitely often as additional terms are added. Then, since it changes at least 20 times, and since each change increases the sum by at least 0.1, after 20 changes in the tenths' place the sum will be increased by at least $20 \times 0.1 = 2$; and thus the partial sums will be at least 3, which is again impossible. Therefore the tenths' place eventually remains constant. Actually, the tenths' place remains constant after $k = 4$ terms have been added (see table 3). In a similar way, one can argue that any digit's location must eventually remain constant. But this argument says that for a given display m, eventually the display is constant. Table 3 shows how the display is constant for $m = 3$ after $k = 6$ terms have been added. Therefore the series $\Sigma 1/k!$ converges. This argument holds in general and shows that a series whose terms are nonnegative converges if the partial sums are bounded. It is easy to go from this result to the various comparison tests in the traditional manner (Apostol 1974, p. 190; Johnson, Kiokemeister, and Wolk 1974, pp. 430–31).

Two Examples

The examples of the previous sections either diverged rather obviously or converged so rapidly that we could observe the behavior of the partial sums after a few terms were added. In this section we shall consider two series in which the question of convergence is more subtle.

Several partial sums for the series $\Sigma 1/k$ and $\Sigma 1/k^2$ are computed in table 4. The partial sums given in the table are not eventually constant for either series. However, we do see that the series $\Sigma 1/k$ grows more rapidly than the series $\Sigma 1/k^2$ and, indeed, the partial sums for $\Sigma 1/k$ have surpassed 6 by the time 300 terms of the series have

TABLE 4

Partial Sums for Two Series

	$s_k = \sum_{i=1}^{k} 1/i$	$s_k = \sum_{i=1}^{k} 1/i^2$
$k = 1$	1.000000	1.000000
$k = 2$	1.500000	1.250000
$k = 3$	1.833333	1.361111
$k = 4$	2.083333	1.423611
$k = 5$	2.283333	1.463611
$k = 6$	2.449999	1.491388
$k = 7$	2.592857	1.511797
$k = 8$	2.717857	1.527422
$k = 9$	2.828968	1.539767
$k = 10$	2.928968	1.549767
$k = 20$	3.597739	1.596163
$k = 30$	3.994987	1.612150
$k = 40$	4.278543	1.620243
$k = 50$	4.499205	1.625132
$k = 100$	5.187377	1.634983
$k = 200$	5.878030	1.639946
$k = 300$	6.282663	1.641606

been added. On the other hand, the partial sums for $\Sigma 1/k^2$ seem to be getting close to a number near 1.6. We might conjecture that $\Sigma 1/k$ diverges and $\Sigma 1/k^2$ converges. How can we decide if this conclusion is valid?

To answer this question we have to look quite carefully at the series. We shall make comparisons as in the previous section.

To show that $\Sigma 1/k$ diverges, we must show that the partial sums are unbounded. One way to do this is to consider what happens to the partial sum s_n when n terms are added to it. For s_1, s_2, s_4, and s_8 we obtain

$$s_1 + \frac{1}{2} \leq s_2$$

$$s_2 + \frac{1}{2} = s_2 + \frac{1}{4} + \frac{1}{4} \leq s_2 + \frac{1}{3} + \frac{1}{4} = s_4$$

$$s_4 + \frac{1}{2} = s_4 + \frac{1}{8} + \frac{1}{8} + \frac{1}{8} + \frac{1}{8}$$

$$\leq s_4 + \frac{1}{5} + \frac{1}{6} + \frac{1}{7} + \frac{1}{8} = s_8.$$

In general, we add $1/(2^n + 1) + 1/(2^n + 2) + \cdots + 1/2^{n+1}$ to s_{2^n} to obtain $s_{2^{n+1}}$. Since

$$\frac{1}{2} = 2^n \cdot \frac{1}{2^{n+1}}$$

$$= \frac{1}{2^{n+1}} + \frac{1}{2^{n+1}} + \cdots + \frac{1}{2^{n+1}}$$

$$\leq \frac{1}{2^n + 1} + \frac{1}{2^n + 2} + \cdots + \frac{1}{2^{n+1}}$$

we increase the value by at least $\frac{1}{2}$ when we replace s_{2^n} by $s_{2^{n+1}}$. Thus, in replacing s_1 by s_{2^n} we increase the sum by at least $n \cdot \frac{1}{2} = \frac{n}{2}$.

Since $s_1 = 1$, it follows that

$$\text{(1)} \qquad 1 + \frac{n}{2} \leq s_{2^n}.$$

This inequality shows that the partial sums are unbounded, since $1 + \frac{n}{2}$ becomes arbitrarily large as n becomes large. Thus $\Sigma 1/k$ diverges.

To make s_{2^n} as large as M it would suffice to take $1 + \frac{n}{2} \geq M$ or, equivalently, $n \geq 2(M - 1)$. Thus $s_{2^{100}}$ is greater than 51, since $100 \geq 2(51 - 1)$. Since 2^{100} is approximately 1.26765×10^{30}, based on this calculation we would have to sum a very large number of terms to be sure that the partial sum would be larger than 51. Fewer terms, although still an enormous number, actually suffice. A challenging problem would be to find the smallest n for which $s_n \geq 51$.

Now consider $\Sigma 1/k^2$. Using arguments like those above, we find that

$$s_3 = s_1 + \frac{1}{2^2} + \frac{1}{3^2}$$

$$\leq s_1 + \frac{1}{2^2} + \frac{1}{2^2} = 1 + \frac{1}{2}$$

$$s_7 = s_3 + \frac{1}{4^2} + \frac{1}{5^2} + \frac{1}{6^2} + \frac{1}{7^2}$$

$$\leq s_3 + \frac{1}{4^2} + \frac{1}{4^2} + \frac{1}{4^2} + \frac{1}{4^2}$$

$$= s_3 + \frac{1}{4} \leq 1 + \frac{1}{2} + \frac{1}{4}$$

$$s_{15} = s_7 + \frac{1}{8^2} + \cdots + \frac{1}{15^2}$$

$$\leq s_7 + \frac{1}{8^2} + \cdots + \frac{1}{8^2}$$

$$= s_7 + \frac{1}{8} \leq 1 + \frac{1}{2} + \frac{1}{4} + \frac{1}{8}.$$

In general, we have

$$s_{2^{n+1}-1} \leq \sum_{k=0}^{n} \frac{1}{2^k} = \frac{1 - \frac{1}{2^{n+1}}}{1 - \frac{1}{2}} \leq \frac{1}{1 - \frac{1}{2}} = 2.$$

Thus the partial sums are bounded by 2 and the series $\Sigma 1/k^2$ converges.

The reason that the partial sums for $\Sigma 1/k^2$ given in table 4 do not stabilize is that even for fairly large k, $1/k^2$ is relatively large and so affects the sum. For example, for $k = 300$, $1/300^2 = .0000111\ldots$ One way to circumvent this difficulty is to replace the series $\Sigma 1/k^2$ by a series that has the same sum but converges more rapidly. This may be done by using the identity

$$(2) \qquad \sum_{k=1}^{n} a_k = \sum_{k=1}^{n-1} k(a_k - a_{k+1}) + na_n.$$

Taking $a_k = 1/k^2$, we obtain

$$\sum_{k=1}^{n} 1/k^2 = \sum_{k=1}^{n-1} (2/(k+1)^2 + 1/k(k+1)^2) + 1/n$$
$$= \sum_{k=1}^{n} 2/k^2 + \sum_{k=1}^{n-1} (1/k(k+1)^2) + 1/n - 2,$$

which may be rewritten

$$(3) \qquad \sum_{k=1}^{n} 1/k^2 = (2 - 1/n) - \sum_{k=1}^{n-1} 1/k(k+1)^2.$$

Repeating this procedure taking $a_k = 1/k(k+1)^2$, we obtain

$$\sum_{k=1}^{n} 1/k(k+1)^2 = (3/8 - 1/(n+1)^2) - (1/2)\sum_{k=1}^{n-1} 1/(k+1)^2(k+2)^2.$$

Combining the preceding equation with equation (3), we obtain

$$\sum_{k=1}^{n} 1/k^2 = 13/8 + (1/2)\sum_{k=1}^{n-2} 1/(k+1)^2(k+2)^2 + (1/n^2 - 1/n).$$

For large n, $1/n^2 - 1/n$ is small and hence $\Sigma 1/k^2$ and $13/8 + (1/2)\Sigma 1/(k+1)^2(k+2)^2$ have the same sum. Some partial sums for $13/8 + (1/2)\Sigma 1/(k+1)^2(k+2)^2$ are given in table 5. Notice that the values of the partial sums s_{200} and s_{300} are identical. It can be shown that this value never changes, and so $\Sigma 1/k^2$ (or $13/8+(1/2)\Sigma 1/(k+1)^2(k+2)^2$) has sum 1.644934 correct to six places. The value correct to nine places can be shown to be 1.644934066.

TABLE 5

Values of the Partial Sum

$$s_k = 13/8 + (1/2)\sum_{i=1}^{k} 1/(i+1)^2(i+2)^2$$

	s_k
$k = 1$	1.638888
$k = 2$	1.642361
$k = 3$	1.643611
$k = 4$	1.644166
$k = 5$	1.644450
$k = 6$	1.644609
$k = 7$	1.644706
$k = 8$	1.644767
$k = 9$	1.644809
$k = 10$	1.644837
$k = 20$	1.644918
$k = 30$	1.644928
$k = 40$	1.644931
$k = 50$	1.644932
$k = 100$	1.644933
$k = 200$	1.644934
$k = 300$	1.644934

An interesting exercise would be to take $a_k = 1/(k+1)^2(k+2)^2$ in (2) to produce an even more rapidly convergent series. To get some idea how many terms would be needed to produce accuracy to six places, several partial sums of the series should be computed.

The idea used above of grouping 2^n terms together can be used to derive a general convergence test sometimes referred to as the Cauchy condensation test, or the 2^n-test. For a discussion see Porter (1972). See Shohat (1933) on the replacement of a slowly converging infinite series with a more rapidly converging series.

Conclusions

A computer or calculator shows how numbers operate when represented as digits rather than as abstract symbols on the pages of a mathematics text. In a sense these numbers are more real than the symbols, and undoubtedly students have had more experience with the digital representations. The approach presented in this paper affords an opportunity to introduce topics some of which have previously been considered advanced. It also shows how machines, instead of being regarded as mere supplementary tools, can be intimately woven into the theoretical framework itself.

REFERENCES

Apostol, T. M. *Mathematical Analysis*. 2d ed. Reading, Mass.: Addison-Wesley Publishing Co., 1974.

Johnson, R. E., F. L. Kiokemeister, and E. S. Wolk. *Calculus with Analytic Geometry*. 5th ed. Boston: Allyn & Bacon, 1974.

Porter, G. J. "An Alternative to the Integral Test for Infinite Series." *American Mathematical Monthly* 79 (June–July 1972): 634–35.

Shohat, J. A. "On a Certain Transformation of Infinite Series." *American Mathematical Monthly* 40 (April 1933): 226–29.

DERIVATIVES ON THE HAND-HELD CALCULATOR

Are you depending on a convincing demonstration of simple limits using a pocket calculator? Caution is advised.

By RAYMOND V. MORGAN and TONY T. WARNOCK

**Sul Ross State University
Alpine, TX 79830**

Current trends are encouraging the use of hand-held electronic calculators as aids to instruction and computation at both the elementary and secondary level as well as in colleges and universities. The analysis of error can no longer be relegated to courses in numerical analysis at the undergraduate level. Many instructors are using calculators to illustrate concepts involving limits and differentiation. The purpose of this article is to point out some of the problems that arise in the use of calculators to illustrate derivatives and to give some illustrations of numerical differentiation techniques. These problems are somewhat explained in the articles by Goldberg and Staib in the Reader Reactions section of the October 1976 issue of the *Mathematics Teacher.*

The results will be recorded using two types of error. The first will be absolute error, or error that is given as

$$|\text{approximation} - \text{true value}|.$$

The second type of error is relative error, which is given as

$$\frac{\text{absolute error}}{|\text{true value}|}.$$

Each type of error is important to some calculations and unimportant to others. In many scientific applications such as surveying and in most engineering problems, the relative error is more useful or more readily available than absolute error. In other applications such as finance or tolerance in engine assembly, absolute error is of greater importance. Usually if calculations involve "large" numbers the relative error is more significant, and with "small" numbers the absolute error is more important. Smith points out that the floating-point number system used on most calculators tends to emphasize relative error rather than absolute error, and so do most engineering and scientific applications.

Some of the problems that arise in using calculators, which can induce error in the most significant digits, are the subtraction of numbers that are very nearly the same size and are rounded or truncated by the calculator. For example, consider 2.648 83 − 2.644 56 = 0.004 27. Round the result to three significant digits 4.27×10^{-3}. If the numbers are first rounded to three significant digits and then subtracted, the result is 2.65 − 2.64 = 0.01. Such problems are inherent in numerical differentiation, since by definition

$$f'(x_0) = \lim_{\delta \to 0} \frac{f(x_0 + \delta) - f(x)}{\delta}.$$

Theoretically the nearer δ is to zero the better approximation

$$\frac{f(x_0 + \delta) - f(x)}{\delta}$$

is to $f'(x_0)$. In using calculators, this is not the case. In fact, for many problems $10^{-5} < \delta < 10^{-3}$ will give the optimum approximation to $f'(x_0)$.

From Taylor's theorem,

$$f(x + \delta) = f(x) + \delta f'(x) + \frac{\delta^2}{2!} f''(\xi) \quad \text{for } \xi \epsilon (x, x + \delta).$$

Thus

$$f(x + \delta) - f(x) = \delta f'(x) + \frac{\delta^2}{2!} f''(\xi),$$

and so

$$\frac{f(x+\delta)-f(x)}{\delta} = f'(x) + \frac{\delta}{2!} f''(\xi),$$

and the error in $\frac{f(x+\delta)-f(x)}{\delta}$ as an approximation to $f'(x)$ is of the order of δ.

A somewhat better approximation can be obtained by using

$$\frac{f(x+\delta)-f(x-\delta)}{2\delta},$$

which is obtained from Taylor's expansion of $f(x+\delta)$ and $f(x-\delta)$. Then

$$\frac{f(x+\delta)-f(x-\delta)}{2\delta} - f'(x) = \frac{\delta^2}{3!} f^{(3)}(\xi)$$

where $\xi \epsilon (x-\delta, x+\delta)$ and the error is of order δ^2. This is a symmetric difference, and if we consider

$$\lim_{\delta \to 0} \frac{f(x+\delta)-f(x-\delta)}{2\delta}$$

we obtain a function known as the Schwartz derivative. It is interesting to note that functions such as $|x|$ have a Schwartz derivative at zero; therefore, caution must be exercised in using this approximation.

Since it is frequently necessary to know the second derivative, Taylor's expansion can be used to obtain the following numerical form with error estimate.

$$\frac{f(x+\delta)+f(x-\delta)-2f(x)}{\delta^2} - f''(x) = \frac{2\delta^2}{4!} f^{(4)}(\xi)$$

for some $\xi \epsilon (x-\delta, x+\delta)$. The error is of order δ^2.

More caution must be used in selecting δ for the second derivative than for the first, because of the presence of δ^2 in the denominator. For example, if $f(x) = x^2$, $f'(x) = 2x$, $f''(x) = 2$, and $f^{(3)}(x) = 0$, so that from the equation above,

$$\frac{f(x+\delta)+f(x-\delta)-2f(x)}{\delta^2} - f''(x) = 0\cdot$$

However, if $\delta = 10^{-6}$ on a 10-digit display calculator, we obtain an answer of one and, if $\delta \leq 10^{-7}$, an answer of zero.

To illustrate these comments, consider the following examples. The first three are typical functions encountered in a beginning calculus course. The fourth is a typical exponential growth function. The fifth example represents a state equation for gas at detonation. The calculating for the last example is performed on three calculators: an eight-digit display Radio Shack model EC-450, an HP45 with ten-digit display, and a TI SR52, which carries twelve digits and displays ten. The Radio Shack calculator performs scientific calculation with six-digit accuracy. All other examples were done on the SR52. The purpose of three machines is to illustrate the importance of extra digits in accuracy when computing derivatives. In each of these examples f_1' is defined by

$$\frac{f(x+\delta)-f(x)}{\delta},$$

and f_2' by

$$\frac{f(x+\delta)-f(x-\delta)}{2\delta}.$$

Also, we use the FORTRAN notation to denote powers of 10. For example, 1E-6 = 10^{-6}. In each example we denote absolute error by "abs" and relative error by "rel." The asterisks in the absf'' and the relf'' columns indicate that the values continue to increase.

Example 1:

$$f(x) = x^5 + x + 1;$$

$$f'(x) = 5x^4 + 1;\ f'(2) = 81$$

$$f''(x) = 20x^3;\ f''(2) = 160$$

If $\delta = 10^{-1}$ and $x = 2$, then

$$f'_1 = \frac{f(2.1)-f(2)}{0.1} = \frac{43.941\,01 - 35}{0.1} = 89.4101$$

and

$$f'(2) = 81.$$

Thus

$$\text{abs}f_1' = |\text{approximation} - \text{true value}|$$

$$= |89.4101 - 81| = |8.4101| = 8.41\text{ E}0,$$

and

$$\text{rel}f'_1 = \frac{\text{absolute error}}{\text{true value}} = \frac{8.4101}{81}$$

$$= 0.103\ 828\ 395\ 1 = 1.04\ \text{E-1}.$$

The other values are calculated in the same manner and are recorded in table 1.

It is interesting to notice that in evaluating the derivatives, more accurate results are obtained if x^5 is evaluated as $(x^2)^2 \cdot x$, instead of evaluating it as x^5 as was done in the example.

Example 2 (see table 2):

$f(x) = \sin x$

$f'(x) = \cos x;\ f'(1) = 0.540\ 302\ 305\ 869$

$f''(x) = \sin x;\ f''(1) = -0.841\ 470\ 984\ 808$

Example 3 (see table 3):

$f(x) = \ln x$

$f'(x) = 1/x;\ f'(3) = 0.333\ 333\ 333\ 333$

$f''(x) = -1/x^2;\ f''(3) = -0.111\ 111\ 111\ 111$

Example 4 (see table 4):

$f(x) = 5000e^{(0.3466)t}$

Typical exponential growth problem

$f'(x) = 1733e^{0.3466t};$

$f'(4) = 6.932\ 732\ 327\ 37\ \text{E}\ 3$

$f''(x) = 600.6578e^{0.3466t};$

$f''(4) = 2.402\ 885\ 024\ 66\ \text{E}\ 3$

Example 5 (see table 5):

$f(x) = x^{-1/4}\exp(x^{-1/4})$

$f'(x) = (-1/4)(x^{-5/4} + x^{-3/2})\exp(x^{-1/4});$

$f'(2) = 0.448\ 619\ 905\ 060$

$f''(x) = (1/16)\ (5x^{-9/4} + 7x^{-10/4} + x^{-11/4})\ \exp(x^{-1/4});$

$f''(2) = 0.353\ 158\ 232\ 389$

Example 6a (see table 6a):

$f(x) = e^{x^2}$

$f'(x) = 2xe^{x^2};\ f'(3) = 4.861\ 847\ 4\ \text{E}\ 4$

$f''(x) = (2 + 4x^2)e^{x^2};\ f''(3) = 3.079\ 17\ \text{E}\ 5$

TABLE 1

δ	absf'_1	relf'_1	absf'_2	relf'_2	absf''	relf''
1E-1	8.41 E 0	1.04 E-1	4.00 E-1	4.94 E-3	2.00 E-2	1.25 E-3
1E-2	8.04 E-1	9.93 E-3	4.00 E-3	4.94 E-5	1.25 E-5	2.00 E-3
1E-3	8.00 E-2	9.88 E-4	4.00 E-5	4.94 E-7	0	0
1E-4	8.00 E-3	9.88 E-5	5.00 E-7	6.17 E-9	1.00 E-2	6.25 E-5
1E-5	7.90 E-4	9.75 E-6	0	0	2.00 E 0	1.25 E-2
1E-6	1.00 E-4	1.23 E-6	0	0	4.00 E 1	2.50 E-1
1E-7	1.00 E-3	1.23 E-5	5.00 E-4	6.17 E-6	1.02 E 4	6.35 E 1
1E-8	0	0	0	0	1.60 E 2	1.00 E 0
1E-9	1.00 E-1	1.23 E-3	0	0	2.00 E 8	1.25 E 6
1E-10	1.00 E 0	1.23 E-2	5.00 E-1	6.17 E-3	*	*
1E-11	1.10 E 1	1.36 E-1	1.00 E 0	1.23 E-2	*	*

TABLE 2

δ	absf'_1	relf'_1	absf'_2	relf'_2	absf''	relf''
1E-1	4.29 E-2	7.95 E-2	9.00 E-4	1.67 E-3	7.01 E-4	8.33 E-4
1E-2	4.22 E-3	7.80 E-3	9.01 E-6	1.67 E-5	7.03 E-6	8.36 E-6
1E-3	4.21 E-4	7.79 E-4	9.03 E-8	1.67 E-7	1.78 E-6	2.12 E-6
1E-4	4.21 E-5	7.78 E-5	6.13 E-9	1.13 E-8	2.31 E-4	2.75 E-4
1E-5	4.39 E-6	8.02 E-6	1.41 E-7	2.61 E-7	8.47 E-3	1.01 E-2
1E-6	1.61 E-6	2.97 E-6	4.06 E-7	7.51 E-7	9.59 E-1	1.14 E 0
1E-7	4.69 E-6	8.69 E-6	1.19 E-6	2.21 E-6	1.31 E-2	1.55 E 2
1E-8	3.23 E-5	5.98 E-5	7.23 E-5	1.34 E-4	*	*
1E-9	6.02 E-4	1.11 E-3	1.00 E-3	1.86 E-3	*	*
1E-10	1.33 E-2	2.46 E-2	5.30 E-3	9.81 E-3	*	*

TABLE 3

δ	absf'_1	relf'_1	absf'_2	relf'_2	absf''	relf''
1E-1	5.44 E-3	1.63 E-2	1.24 E-4	3.71 E-4	6.18 E-5	5.56 E-4
1E-2	5.54 E-4	1.66 E-3	1.23 E-6	3.70 E-6	4.69 E-7	4.22 E-6
1E-3	5.56 E-5	1.67 E-4	8.67 E-9	2.60 E-8	7.11 E-6	6.40 E-5
1E-4	5.61 E-6	1.68 E-5	2.83 E-8	8.50 E-8	1.01 E-3	9.10 E-3
1E-5	1.13 E-6	3.40 E-6	2.83 E-7	8.50 E-7	1.01 E-1	9.10 E-1
1E-6	1.33 E-6	4.00 E-6	3.33 E-7	1.00 E-7	14.11 E0	1.27 E 2
1E-7	1.33 E-5	4.00 E-5	8.33 E-6	2.50 E-5	1.50 E 3	1.35 E 4
1E-8	1.33 E-4	4.00 E-4	8.33 E-5	2.50 E-4	*	*
1E-9	1.33 E-3	4.00 E-3	8.33 E-4	2.50 E-3	*	*
1E-10	1.33 E-2	4.00 E-2	8.33 E-3	2.50 E-2	*	*
1E-11	1.33 E-1	4.00 E-1	8.33 E-2	2.50 E-1	*	*

TABLE 4

δ	absf'_1	relf'_1	absf'_2	relf'_2	absf''	relf''
1E-1	1.21 E 2	1.75 E-2	1.39 E 0	2.00 E-4	2.41 E-1	1.00 E-4
1E-2	1.20 E 1	1.73 E-3	1.39 E-2	2.00 E-6	3.48 E-3	1.45 E-6
1E-3	1.20 E 0	1.73 E-4	1.48 E-4	2.12 E-8	1.65 E-1	6.87 E-5
1E-4	1.20 E-1	1.73 E-5	7.74 E-5	1.12 E-8	1.21 E 1	5.04 E-3
1E-5	1.26 E-2	1.83 E-6	1.73 E-4	2.49 E-8	1.10 E 3	4.57 E-1
1E-6	1.77 E-2	2.55 E-6	7.33 E-3	1.06 E-6	1.48 E 5	6.14 E 1
1E-7	2.32 E-1	3.35 E-5	2.32 E-1	3.35 E-5	1.00 E 7	4.16 E 3
1E-8	2.26 E 0	3.27 E-4	2.32 E-1	3.35 E-5	1.50 E 9	6.24 E 5
1E-9	1.73 E 1	2.49 E-3	7.73 E 0	1.12 E-3	1.50 E11	6.24 E 7
1E-10	4.33 E 2	6.24 E-2	1.83 E 2	2.64 E-2	5.00 E13	2.08 E 9
1E-11	1.93 E 3	2.97 E-1	1.93 E 3	2.79 E-1	1.00 E15	4.16 E11

TABLE 5

δ	absf'_1	relf'_1	absf'_2	relf'_2	absf''	relf''
1E-1	1.69 E-2	3.77 E-2	7.72 E-4	1.72 E-3	7.04 E-4	1.99 E-3
1E-2	1.76 E-3	3.92 E-3	7.70 E-6	1.72 E-5	7.16 E-6	2.03 E-5
1E-3	1.76 E-4	3.93 E-4	7.89 E-8	1.76 E-7	9.77 E-6	2.77 E-5
1E-4	1.75 E-5	3.90 E-5	4.99 E-8	1.11 E-7	5.82 E-5	1.65 E-4
1E-5	1.01 E-6	2.24 E-6	2.95 E-7	6.57 E-7	8.68 E-2	2.46 E-1
1E-6	9.09 E-6	2.03 E-5	2.09 E-6	4.67 E-6	3.65 E 0	1.03 E 1
1E-7	7.01 E-5	1.56 E-4	3.51 E-5	7.82 E-5	1.10 E 3	3.11 E 3
1E-8	2.80 E-4	6.24 E-4	1.70 E-4	3.79 E-4	9.00 E 5	2.55 E 5
1E-9	1.04 E-2	2.31 E-2	3.38 E-3	7.53 E-3	*	*
1E-10	1.41 E-1	3.15 E-1	6.14 E-2	1.37 E-1	*	*
1E-11	1.45 E0	3.24 E0	6.51 E-1	1.45 E0	*	*

TABLE 6a

δ	absf'_1	relf'_1	absf'_2	relf'_2	absf''	relf''
1E-1	1.95 E 4	4.00 E-1	5.21 E 4	1.07 E 0	1.20 E 4	3.88 E-2
1E-2	1.57 E 3	3.24 E-2	3.38 E 1	6.97 E-4	1.13 E 2	3.67 E-4
1E-3	1.54 E 2	3.16 E-3	5.26 E-1	1.08 E-5	1.92 E 3	6.23 E-3
1E-4	1.15 E 1	2.37 E-4	1.53 E 0	3.13 E-5	*	*
1E-5	8.15 E 1	1.68 E-3	3.15 E 1	6.48 E-4	*	*
1E-6	1.83 E 3	3.80 E-2	3.81 E 2	7.85 E-3	*	*
1E-7	8.62 E 3	1.77 E-1	1.38 E 3	2.84 E-2	*	*

Example 6b (see table 6b):

$f(x) = e^{x^2}$

$f'(x) = 2xe^{x^2}$; $f'(3) = 4.861\ 850\ 356$ E 4

$f''(x) = 2\ e^{x^2}(2x^2 + 1)$;

$f''\ (3) = 3.079\ 171\ 892$ E 5

Example 6c (see table 6c):

$f(x) = e^{x^2}$

$f'(x) = 2xe^{x^2}$; $f'(3) = 4.861\ 850\ 356\ 54$ E 4

$f''(x) = (2 + 4x^2)e^{x^2}$;

$f''(3) = 3.079\ 171\ 892\ 47$ E 5

These examples illustrate how the theoretical properties of derivatives are altered when computations are performed on hand-held electronic calculators or on large computers if there is not sufficient digit capacity. Example 6 illustrates the marked difference in using 8-, 10-, or 12-digit calculators. Thus, if a calculator is used to illustrate differentiation, one should indicate that limited display is responsible for the failure of the difference quotient to converge as δ decreases.

If derivatives must be calculated numerically, the symmetric difference

$$\frac{f(x + \delta) - f(x - \delta)}{2\delta}$$

converges more rapidly than does the one-sided definition, since it has an error term of order δ^2. Hildebrand points out that the smaller δ is chosen to be, the smaller the truncation error will be, where truncation error is defined to be any error not related to round-off error. However, a decrease in δ induces a possible increase in calculator round-off error. Conversely, a reduction in round-off error would generally lead to an increase in truncation error. Figure 1 illustrates this graphically.

Thus, δ should be selected so that both errors are as small as possible. He gives the optimal value of δ as $\delta_{opt} \approx 1.8\epsilon^{1/3}M^{-1/3}$, where $|f^{(3)}(x)| < M$ on $[x_0 - \delta, x_0 + \delta]$ and ϵ is the maximum possible error in $f(x_0)$. If a hand calculator uses 12 digits for calculation and rounds back to 10 in the display, then the maximum possible error in $f(x_0)$ is

TABLE 6b

δ	absf'_1	relf'_1	absf'_2	relf'_2	absf''	relf''
1E-1	1.95 E 4	4.01 E-1	3.48 E 3	7.18 E-2	1.20 E 4	3.88 E-2
1E-2	1.57 E 3	3.24 E-2	3.40 E 1	7.00 E-4	1.18 E 2	3.82 E-4
1E-3	1.54 E 2	3.17 E-3	3.40 E-1	7.00 E-6	2.81 E 0	9.13 E-6
1E-4	1.54 E 1	3.17 E-4	1.44 E-3	2.96 E-8	8.28 E 1	2.69 E-4
1E-5	1.50 E 0	3.08 E-5	3.56 E-3	7.32 E-8	7.29 E 3	2.57 E-2
1E-6	6.04 E-1	1.04 E-5	5.04 E-1	1.04 E-5	*	*
1E-7	1.49 E 0	3.08 E-5	1.50 E-0	3.08 E-5	*	*
1E-8	1.85 E 1	3.81 E-4	1.85 E 1	3.81 E-4	*	*
1E-9	3.81 E 2	7.85 E-3	3.81 E 2	7.85 E-3	*	*

TABLE 6c

δ	absf'_1	relf'_1	absf'_2	relf'_2	absf''	relf''
1E-1	1.95 E 4	4.01 E-1	3.48 E 3	7.18 E-2	1.20 E 4	3.88 E-2
1E-2	1.57 E 3	3.24 E-2	3.40 E 1	7.00 E-4	1.18 E 2	3.82 E-4
1E-3	1.54 E 2	3.17 E-3	3.40 E-1	7.00 E-6	1.18 E 0	3.84 E-6
1E-4	1.54 E 1	3.17 E-4	3.37 E-3	6.94 E-8	1.08 E-2	3.49 E-8
1E-5	1.54 E 0	3.17 E-5	2.15 E-4	4.43 E-9	1.28 E 1	4.16 E-5
1E-6	1.43 E-1	3.16 E-6	5.65 E-4	1.16 E-8	6.08 E 3	1.98 E-2
1E-7	3.36 E-2	6.90 E-7	3.86 E-2	7.93 E-7	3.92 E 5	1.27 E 0
1E-8	8.04 E-1	1.65 E-5	4.04 E-1	8.30 E-6	2.03 E 7	6.59 E 1
1E-9	1.50 E 0	3.09 E-5	1.00 E 0	2.06 E-5	*	*
1E-10	4.85 E 1	9.98 E-4	2.35 E 1	4.83 E-4	*	*
1E-11	9.19 E 2	1.89 E-2	4.68 E 2	9.64 E-3	*	*

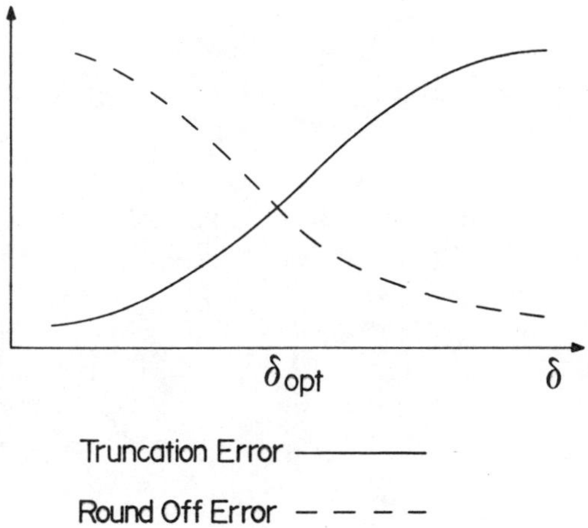

Fig. 1

$\epsilon = 5 \times 10^{-13}$. Then $\delta_{opt} \approx 1.8(5 \times 10^{-13})^{1/3}M^{-1/3} = 1.43 \times 10^{-4}M^{-1/3}$. Thus, if M is near unity, the optimal choice for δ is between 10^{-4} and 10^{-5}. Similar calculations give the appropriate choice δ for calculators, which display fewer digits.

This article has illustrated that in theory, better approximations are obtained by decreasing δ; in practice, however, this is not necessarily true. In fact, in many instances, errors are greatly increased by choosing δ too small. Such problems should be adequately explained to students to whom limits are illustrated using calculators. In most instances, the two-sided derivative illustrates convergence in greater detail. The rounding error is the same order of magnitude as the one-sided form, but the truncation error is one order better.

BIBLIOGRAPHY

Hildebrand, F. B. *Introduction to Numerical Analysis,* 2d ed. New York: McGraw-Hill, 1974.

Smith, Jon M. *Scientific Analysis on the Pocket Calculator.* New York: John Wiley & Sons, 1975.

Research

Numerous studies on the hand calculator have been conducted during the past year or two, but very little has been reported in the journals. Additional research is still needed.

The *Mathematics Teacher* Editorial Panel developed a questionnaire on computational skill. Responses from teachers, mathematicians, and laypersons are given in the first article of this chapter. The impetus for the study came from the expectation that calculators would be available to nearly everyone in the United States in the near future.

Schnur and Lang conducted an experimental study with sixty youngsters and found significant growth in achievement among students using calculators compared to those not using them.

The trial project reported by Sullivan was referred to by Hawthorne in his article in chapter one. The teachers involved observed that the hand-held calculator serves as a motivational device and encourages students to investigate topics that they wouldn't ordinarily tackle.

The final article reports the findings of a study conducted with 600 seventh-grade students over a one-year period. Rudnick and Krulik draw some conclusions based on student achievement and a questionnaire survey of parents.

WHERE DO YOU STAND?

COMPUTATIONAL SKILL IS PASSÉ

SOON nearly everyone in the United States who faces an arithmetic problem will be able to call on a low-cost electronic calculator as an aid. For more elaborate calculation, remote communication with a computer will be almost as easy. This development has led many teachers and students to question the high instructional and testing priority currently assigned to speed and accuracy in arithmetic computation. Their doubts have been expressed in a variety of controversial propositions and proposals for curricular change.

The MATHEMATICS TEACHER Editorial Panel posed the following seven issues to a sample of teachers, mathematicians, and laymen. Their responses are given here in percentage form, along with some of their positions and justifications, which help to identify the consequences of emerging technology and alternative instructional policies.

1. Facility with arithmetic computation is the major goal of elementary and junior high school mathematics teaching today.

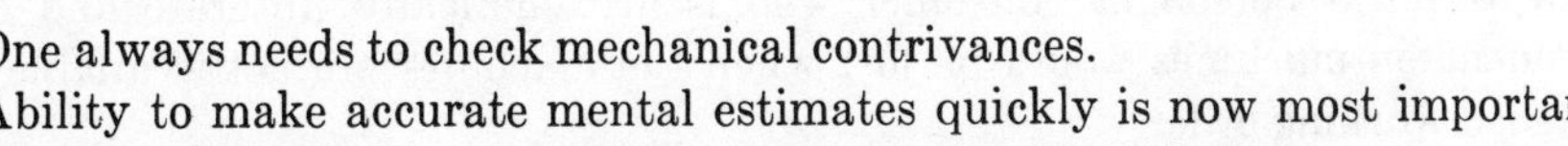

2. Speed and accuracy in arithmetic computation is still essential for a large segment of business and industrial workers and intelligent consumers.

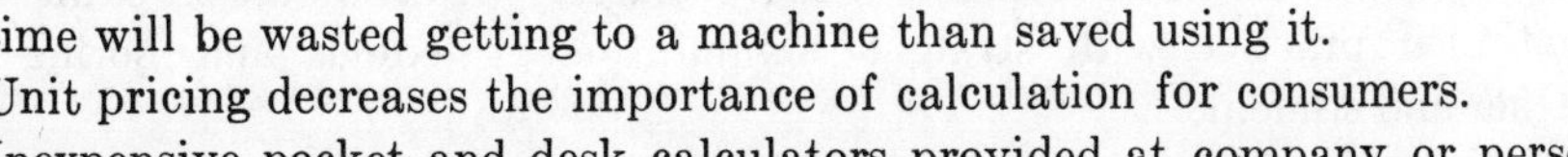

—One always needs to check mechanical contrivances.
—Ability to make accurate mental estimates quickly is now most important.
—One doesn't carry his pocket calculator around at all times.
—Calculators will not soon be as readily available as pencil and paper; more time will be wasted getting to a machine than saved using it.
—Unit pricing decreases the importance of calculation for consumers.
—Inexpensive pocket and desk calculators provided at company or personal expense have all but eliminated the need for computational skill.
—When is the last time you saw a salesperson do mental or paper-and-pencil computation?

3. Impending adoption of metric measurement implies that computation with rational numbers should be largely confined to decimal fractions.

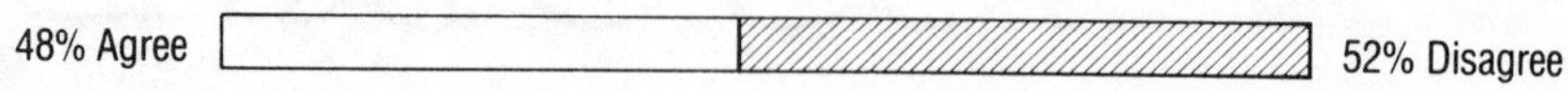

—The housewife will still want fractions of a recipe, and these fractions and ratios are not always best expressed as decimals.
—This is true for the low achievers; others should spend more time with p/q.
—Working with algebraic fractions requires considerable knowledge of all kinds of fractions.
—As long as we measure time to the quarter of an hour, sell shares of stock at $35\frac{7}{8}$, measure material by eighths of yards, or even concern ourselves with degree and radian measure, we must continue to work with all kinds of fractions.
—Probability requires common fractions.
—Halves, thirds, and quarters are too commonly needed.
—Eventually the average citizen may have no use for fractions of the type p/q.
—We need more experience with the metric system before we can express an opinion.
—We should examine the European experience.
—The decimal notation has always made more sense than common fractions; it makes use of the basic whole-number computational algorithms and avoids far-out common denominator problems, and so on.

4. In the face of declining arithmetic computation test scores, the energies of mathematics instruction should be concentrated on these skills until achievement reaches mastery levels.

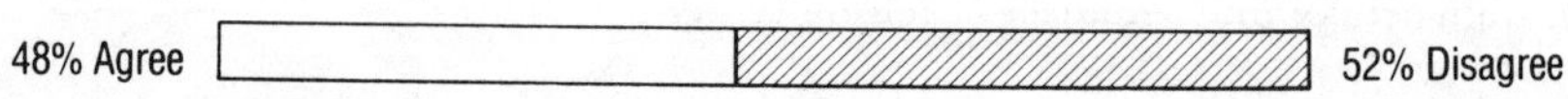

—Mastery of computation is not essential to do the kind of arithmetic thinking that is important in programming, estimating, and checking answers from machine computation.
—Yes, particularly for those teachers who have misunderstood the call for emphasis on concepts and excluded drill.
—The business person or consumer who is arithmetically illiterate and dependent on machines would be lost when the machines are not available or in good working order.
—Kids compute as well now as they ever have.
—Not everyone can be expected to achieve mastery of computation skills.
—Continued practice with straight computation is tedious and boring to teacher and student.
—Introducing geometry into lower grades is as important as computational skills.
—Spend the money on making calculators available so we can get on to more important mathematics, like learning *when* to perform arithmetic operations.
—Mastery level might be a long way off, but we should certainly strive for significant improvement in skills.

—What we need is to establish individual mastery levels for each child, reflecting his or her ability, expectations, and so on.
—The energies of mathematics instruction should be spent on problem-solving skills and applications of mathematics.

5. Weakness in computational skill acts as a significant barrier to learning of mathematical theory and applications.

61% Agree 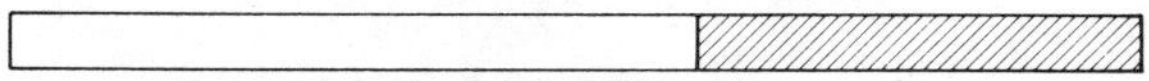 39% Disagree

—Yes, to the extent that errors in calculation prohibit building on that result.
—There is correlation, but doubtful causal relation.
—It is through arithmetical examples that one gets the feel of what theory and applications are about.
—How can one expect to learn algebra as generalized arithmetic without knowing arithmetic?
—If a person is not sure of an algorithm, he cannot concentrate on the theoretical issue or application.
—I know a brilliant mathematician who cannot do basic computation.
—Students today show *increased* knowledge of theory and applications.
—This is true when it comes to applying theory to applications; those who make contributions to theory but have weak computational skills are exceptional.
—What electronic calculator helps factor trinomials?
—Many instructional texts use simple skills in illustrating mathematical theory and applications.
—Many slower students lacking computational skill have, when given access to a calculator, easily mastered some difficult theories and applications.
—The more important understanding of concepts and quantitative relationships can easily be supported by computation equipment if student skills are weak.

6. Every seventh-grade mathematics student should be provided with an electronic calculator for his personal use throughout secondary school.

28% Agree 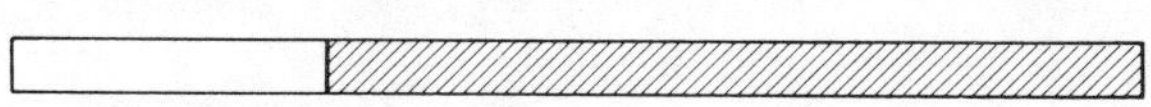 72% Disagree

—Cost would be prohibitive for the initial supply and the replacements due to theft, damage, and so on.
—With costs plummeting downward, a school system can no longer consider the cost prohibitive.
—Students would become too dependent on the calculator as a crutch.
—This should wait until grade 8 or 9 or when students demonstrate computational proficiency without the calculator.
—Maybe even at an earlier age!
—In China, students are taught to use the abacus early.
—Access at least should be provided.
—Students minds would get lazy and operate less efficiently if the machines were available.

7. Availability of calculators will permit treatment of more realistic applications of mathematics, thus increasing student motivation.

96% Agree  4% Disagree

—We won't have to avoid the messy real-life situations or reach for a set of tables; we can deal with the approximations encountered in real measures.
—In my experience, student interest and success increase when calculators are available for use.
—The motivation is often short-lived and artificial.
—The standard text problems rigged with "nice" answers deceive students.
—Calculators will support the efforts of less able students.

How Do You Feel?

Have the above questions identified the key issues in determining the impact of calculating equipment on mathematics education?

Do the opinions of our preliminary sample reflect the beliefs of most NCTM members?

Have you had exceptionally good or bad experiences with calculator use that can be shared profitably with other readers of the MATHEMATICS TEACHER?

The Panel welcomes your reaction.

Just pushing buttons or learning?—a case for minicalculators

JAMES O. SCHNUR and
JERRY W. LANG

This article is tangible evidence of what may happen when an IHE (institution of higher education) person and an LEA (local education agency) person talk and plan together. James Schnur is professor of education and head of the Department of Curriculum and Instruction at the University of Northern Iowa in Cedar Falls. Jerry Lange is principal of the Franklin Elementary School in Muscatine, Iowa.

The minicalculator has captured the imagination and approval of some educators while receiving the disapproval of others. It has already captured the attention of people in many walks of life. It is lauded by a sizeable group within the public as a means through which one can bypass the drudgery of mathematics. Many adults become most appreciative of this pocket-sized device when performing such tasks as completing income tax forms, balancing the checkbook, determining how much carpeting it will take to cover the living room, and countless other tasks.

The minicalculator has quickly become a very affordable and available commodity.

Simple models are now available for as little as ten dollars. They are available not only from business machine establishments but from discount stores, the most popular mail-order houses, and even in the multipurpose drug store.

Despite the availability of the minicalculator and its ever increasing public usage, controversy exists among educators as to its proper usage in the school or whether it should be allowed in the school setting at all. Its advocates refer to it—as an essential implement in the newest mathematics (Higgins 1974)—as a motivating device (Mastbaum 1969)—as a means toward immediate reinforcement of results, a significant learning strategy (Lewis 1974). Among its adversaries, James McKinney, Professor of Mathematics at California Polytechnic State University in Pomona, states the case (in "Great Calculator Debate," 1974) as follows:

> If what we're talking about is reducing tedious calculations, then perhaps minicalculators can be an aid, but teaching a student to push buttons won't help him if what he needs is more instruction in actual addition, subtraction, multiplication and division—I can't think of any reason why a fourth or fifth grader should even

> see one, after all that's when we're trying to teach basic arithmetic. (p. 12)

These differences of opinion along with seeking answers to implementation questions pondered by the investigators provided the stimulus for the study reported here.

The purpose of this investigation was to seek answers to the following questions:

1. Would a group of youngsters of elementary school age achieve greater mathematical computational ability through the controlled usage of the minicalculator than would their noncalculator usage peers, and will such increased ability on the part of minicalculator users transfer to a posttest situation where the minicalculator is not used?
2. Would there be any significant interaction between the sex of the youngster and minicalculator usage or nonusage?
3. Would there be any significant interaction between factors of ethnic/economic background (migrant/nonmigrant status) of the youngster and minicalculator usage or nonusage?

The sample consisted of sixty youngsters enrolled in a summer compensatory education program located in Franklin Elementary School in Muscatine, Iowa. Twenty-five of these youngsters are classified as dependents of migrant agricultural adults. The majority of this group was also bilingual/bicultural of Mexican-American heritage. The remaining youngsters in the sample were regular students from the school system who were seeking remedial assistance.

To answer the research questions youngsters were randomly assigned to either a male or female, migrant or nonmigrant grouping within the calculator usage or noncalculator usage category. Logistically the youngsters received instruction by being randomly assigned into classroom settings that were intergrated in terms of ethnic/economic background (migrancy/nonmigrancy) and sex. These classrooms were four in number. Two were designated as experimental settings and two were designated as control settings. Each classroom also incorporated multiage grouping with age ranges of 9 to 14.

Instruction delivered to all groups was uniform with the exception of calculator usage. All students received instruction in the four basic operations (+, ×, −, ÷) on whole numbers from multiple sources that included the Individualized Computational Skills Program (Houghton-Mifflin Co.) skills sheets and models, System 80 Beginning Math Concept Lessons (Borg-Warner Corp.), Veri Tech Boards (Educational Teaching Aids), Language Master Math Fact Program Cards (Bell and Howell Corp.), and teacher devised learning center activities. The duration of instruction was four weeks.

All youngsters were pretested with Form A of the *Individualized Computational Skills Program Computational Test 3–4* (Houghton-Mifflin Co.) on 3 July 1975. The posttest was Form B of this same instrument and was administered on 5 August 1975. These pre- and posttest scores yielded the statistical analysis data. The pretest scores were also used to prescribe individualized learning activities for all students. Calculators were not used on the pre- or posttests.

The treatment groups shared the use of eight minicalculators that provide a six-digit display that can be expanded to an additional six-digit display at the press of an appropriate button. These calculators perform the four basic arithmetic operations of addition, subtraction, multiplication, and division. They also provide a full floating decimal, a clear-last-entry button and an all-clear button. They were operated with AC adaptors so as to eliminate the problem of discharged batteries. This proved to be a very valuable, low-cost addition. The eight calculators were shared by two experimental groups. This was accomplished by scheduling one mathematics period subsequent to the other. This same procedure of scheduling was followed by the control groups.

The experimental group teachers first instructed their students in the proper use of the calculators. This activity substantiated the feeling of Gaslin (1975) when he stated that most children are easily able to master the operation of calculators. This initial procedure was followed by scheduling each experimental youngster's operation of a minicalculator to a minimum of fifty minutes each week with a time sheet maintained for each student in order to assure compliance with the time guidelines established. These youngsters were taught to use the minicalculators for (1) verifying correctness of written answers on completing assigned computational problems, and (2) computation of actual answers to problems on every third Individualized Computational Skills Program (I.C.S.P.) skill sheet, with the exception of the last skill sheet for each basic computational strand, which was used as a written posttest for each specific skill.

As the investigators personally observed youngsters in the experimental groups at work we were impressed by their desire to operate the minicalculators to both prove their work and to find answers to their computational problems. They appeared to be highly motivated throughout the mathematics instructional period. Such observations substantiate those of others reporting these factors (Gaslin 1975, Hawthorne 1973, Lewis 1974). At the same time the control groups were having their own individualized experiences with mathematics and demonstrated no envy at not having access to the minicalculators.

To seek answers to the research questions stated previously, mean gain scores were generated from the data collected from the *I.C.S.P. Computational Test 3–4*, Form A (pretest) and Form B (posttest). These data were controlled for calculator/noncalculator usage, sex, and migrant/nonmigrant status. The statistical procedure to analyze these data was the analysis of variance (ANOVA) factorial design contained in the *Statistical Package for the Social Sciences* (SPSS)—Version 6.00. This procedure was performed by the University of Northern Iowa Computer Center. The results are presented in table 1.

CONCLUSIONS

With regard to the first research question that sought the answer as to whether the experimental-minicalculator usage groups would gain significantly more whole number computational ability and whether any such gained ability would transfer to a nonminicalculator posttesting situation, the data seem to indicate that this is indeed the case. The growth factor as measured by the nonminicalculator usage posttests for the experimental group when compared to the control group data were significantly improved to the .001 alpha level.

The second question dealing with sex × calculator usage interaction proved to be nonsignificant although a slight trend is detectable (1.291 F score, significant at less than or equal to 0.260). Regarding this point, the slight trend is traceable to a slightly favorable female-calculator usage relationship. While this is not significant it

Table 1

Source	*Sum of squares*	*DF*	*M.S.*	*F*	*Significance of F*	
Total	2823.645	59	47.858			
Sex	98.309	1	98.309	2.458	.119	NS
Calculator usage	505.886	1	505.886	12.649	.001	Significant
Migrant	94.224	1	94.224	2.356	.127	NS
Sex × Calculator	51.614	1	51.614	1.291	.260	NS
Sex × Migrant	6.175	1	6.175	0.154	.999	NS
Calculator × Migrant	19.199	1	19.199	0.480	.999	NS
Sex × Calculator × Migrant	7.555	1	7.555	0.189	.999	NS
Error	2079.631	52	39.993	—	—	—

was interesting for the investigators to observe.

The third question dealing with any interaction between factors of ethnic/economic background (migrant/nonmigrant status) and appropriate gain scores, growth proved to be nonsignificant.

SUMMARY AND IMPLICATIONS

This investigation suggests that the incorporation of a simple minicalculator, as an instructional supplement to an erstwhile standard, individualized remedial mathematics program, can yield significant achievement ability growth that will transfer into nonminicalculator testing situations. This was accomplished over a relatively short period of time (four weeks).

The teachers of the experimental groups required no special training and were able to incorporate minicalculators into their regular instructional routines with relative ease. It does not appear, therefore, that this is an instructional device that demands rigorous inservice training to enhance the probability of positive results.

The cost of the eight minicalculators used in this investigation was certainly not prohibitive. The expenditure of a few hundred dollars for minicalculators that are highly motivational and that provide immediate knowledge of results is a cost comparable to the purchase of other instructional material that may not provide these instructional benefits.

It can also be implied, within the scope of this investigation that improvement traceable to the minicalculator is maintained regardless of the sex of the learner or whether the learner is nonmigrant/migrant, bilingual/bicultural of Mexican-American heritage.

The results of this study seem to suggest that the minicalculator does have a place in the elementary classroom. It is hoped that further investigations will help to answer questions dealing with many more specific applications and implications surrounding its usage in the elementary schools of today and tomorrow. The minicalculator seems to be here to stay—it is up to educators to explore the many facets of its implementation rather than to disclaim or ignore its possible usage.

References

Gaslin, William L. "A Comparison of Achievement and Attitudes of Students Using Conventional or Calculator-Based Algorithms for Operations on Positive Rational Numbers in Ninth-Grade General Mathematics." *Journal for Research in Mathematics Education* 6 (March 1975):95–108.

"Great Calculator Debate." *Nations Schools and Colleges* 1 (December 1974):12–14.

Hawthorne, Frank S. "Hand-Held Calculators: Help or Hindrance?" *Arithmetic Teacher* 20 (December 1973):671–72.

Higgins, Jon L. "Mathematics Programs Are Changing." *Education Digest* 40(December 1974):56–58.

Lewis, Phillip. "Minicalculators Have Maxi-Impact." *Nations Schools* 93:60+.

Mastbaum, Sol. "A Study of the Relative Effectiveness of Electric Calculators or Computational Skills Kits in the Teaching of Seventh Grade Students in Mathematics." Ph.D. dissertation, University of Minnesota, 1969.

Using hand-held calculators in sixth-grade classes

JOHN J. SULLIVAN

A consultant in mathematics education for the New York State Education Department, John Sullivan was codirector of classroom trials of hand-held calculators in 1973–74. Frank Hawthorne was the other codirector.

During the 1973–74 school year, two sixth-grade classes in New York conducted classroom trials of hand-held calculators. Each child in these classes had a Bowmar "Brain" hand-held calculator for his use during mathematics lessons each day. The project was organized by the Bureau of Mathematics Education, New York State Education Department. The calculators were provided free of charge by Bowmar/ALI, Inc., and supervision was provided by the principals of the project schools.

Several ground rules were established. Teachers were careful to make sure that the use of calculators did not cause a loss of computational skill. The calculators were available during lessons but not for tests, and were not taken home. A determined effort was made to inform students and parents about what was taking place.

Secure facilities for storage of the calculators were necessary. In each case a closet was modified for calculator storage by having strips of electrical plugs put in so that the calculators could be recharged. Each calculator had either a large numeral or a child's name fastened to it so that they could be distributed quickly. Sixty-eight calculators were in use during the year; eight of them developed malfunctions and had to be returned to the manufacturer. None were stolen; none were damaged.

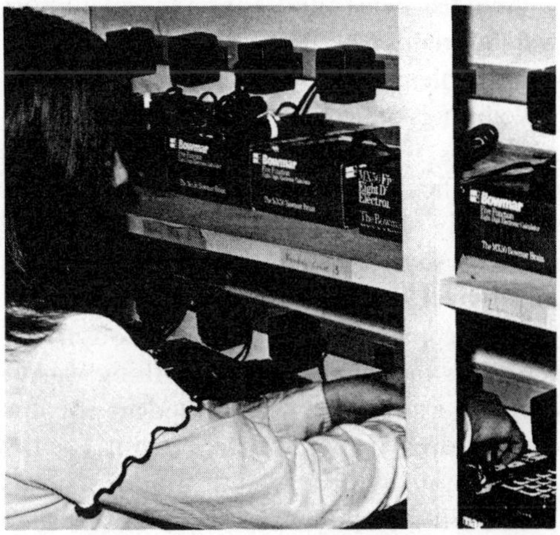

The major goal was to try to discover how (and if) the calculators could enrich, supplement, support, and motivate the regular program. The calculators did catch the interest of the children. A rather high level of interest was maintained throughout the year. It is not possible to measure how much of the positive motivation was due to factors other than the calculator, but observers estimate that the calculators with their amazing speed and accuracy made a definite contribution. Most of the children thought that it was fun to use a calculator. There seemed to be no difficulty in learning how to use the instrument, which has the capacity to add, multiply, divide, and subtract, with floating decimal numeration and a convenient "percent" button. After using the calculator for a time it became evident that many sophisticated mathematical calculations could be done—raising a number to a power or determining factorial numbers, for example.

The calculators were useful for checking answers. A project guideline encouraged their use in the intermediate steps of learning an algorithm. For example, if the lesson objective concerned learning the division algorithm, it was considered appropriate to use the calculator for the intermediate steps involving multiplication and subtraction.

One teacher found that the calculators were extremely helpful in working with verbal problems. His students competed in developing their own complicated verbal problems since the computations offered little difficulty with calculators available. Curiously, the project statistical analysis did not show that the calculators helped improve skill in solving verbal problems.

To the mathematics specialists observing the trials the most heartening thing was to see how the calculators seemed to encourage children to explore topics not usually studied intensively in sixth grade, such as probability, exponents, sequences, prime numbers, palindromes, negative numbers, division by zero, divisibility, and permutations. Many topics from the regular program were supported very well by the calculators—averages, rounding numbers, numeration, factoring, and, of course, the fundamental operations. The machines are well designed for percent problems but this topic does not get much attention in sixth grade.

The following are some of the problems children explored with interest:

1. What is the sum of n consecutive numbers? Can you find a pattern in the answers?

$$1 + 2 = 3$$
$$1 + 2 + 3 = 6$$
$$1 + 2 + 3 + 4 = 10$$

2. What is the total number of possible arrangements of the letters D, R, I, V, E? (Calculators are well suited to computing factorial numbers; e.g., $5 \times 4 \times 3 \times 2 \times 1$.)

3. Is 6713 a prime number? (Children learned a lot about divisibility.)

4. "Quacky quotients." Calculators can be used in games like this one. Each of two players selects a number randomly, and one number is divided by the other. If the first digit in the quotient is 0, 1, 2, 3, or 4, one player scores a point; if the first digit is 5, 6, 7, 8, or 9, the second player scores a point. (This is not a fair game!)

5. By adding a number to a number formed by reversing the digits, a palindrome will be formed in a number of steps. For example:

$$\begin{array}{r} 67 \\ +76 \\ \hline 143 \\ +341 \\ \hline 484 \end{array}$$

How many steps are required to convert 78 to a palindrome?

An important point is illustrated by some of these examples. The hand-held calculator encourages students to investigate topics ordinarily out of bounds because of computational complexity. Some practical topics such as compound interest can be handled nicely by sixth graders using calculators.

The outstanding impact of the calculator may have been its power to motivate increased attention to decimal fractions. Children took an interest in the relationship between common fractions and decimal fractions from the outset. This is an important advantage derived from using calculators.

Mathematics educators have become skeptical about mechanical miracles that will revolutionize instruction. They have been disappointed so many times. The hand-held calculator, however, may be in a unique class. It has powerful mathematical capabilities which teachers may have to recognize, understand, and exploit.

The minicalculator: friend or foe?

JESSE A. RUDNICK
and STEPHEN KRULIK

As professor and coordinator of secondary mathematics education, Jesse Rudnick is involved with the undergraduate and graduate mathematics education programs. Stephen Krulik, as associate professor of mathematics education, teaches graduate courses in mathematics education. He is also responsible for supervising intern teachers in secondary school mathematics.

INTRODUCTION

The age of the minicalculator is now! Not since the printing press has any invention had such potential for revolutionizing education, particularly mathematics education. Following closely on the heels of the "new math" and "metrics," the "minicalculator" is creating greater controversy than either of its predecessors. The multitude of articles that appear in journals and the general press, as well as the number of sessions devoted to the calculator at NCTM meetings and the attendance at these sessions, give strong evidence that teachers, parents, and school administrators are interested in and concerned about its impact on the curriculum, the classroom, and the student.

Uppermost in the minds of most persons, particularly parents, is whether the use of the calculator will impair the students' ability to perform the paper-and-pencil algorithms. Most educators who advocate the use of the calculator qualify their stand by stipulating that it not be used until after the students have, to some degree, mastered the operations. However, there is no evidence that dependence on the machine will not result in forgetting. Forgetting and thus dependence on the machine is the major concern of the opponents of calculators in the classroom.

The issue of the minicalculator and its effects is a controversial one with educators throughout the world. Current literature on this issue abounds with claims and counterclaims. However, in viewing this literature, one fact seems evident: the opinions expressed are not based on solid research. Rather, they are conjectures or the results of small-scale, limited experiments. Indeed a significant portion of the calculator literature deals with its classroom use—this in spite of the lack of any data on its effect.

OVERVIEW OF THE STUDY

In order to shed some light on this important area, the Mathematics Education Action Research Center at Temple University (MEARC) in conjunction with the West Chester, Pennsylvania, public schools, in September 1975, initiated a controlled experiment with 600 seventh-grade students. The experiment, which was to continue for the entire school year, was designed to measure the effect of the availability and use of a minicalculator on the students' total mathematics achievement and their ability to perform pencil-and-paper basic skills.

MEARC staff for this project includes Bruce Burt, West Chester School District; David Kapel, University of Nebraska at Omaha; Nola Blye, Cheney State College; Stephen Krulik, L. Waldo Rich, Jesse Rudnick, Ann M. Wilderman, Temple University.

Students in one half of the seventh-grade classes in two schools were randomly selected as the experimental group while the remaining half served as the control. The experimental group had minicalculators available to them each day for use during their mathematics class period. After a three-day learning period with the calculator, these students were "on their own" as to how and when to use the machine. *No modifications were made in the mathematics curriculum.* Each student kept a log indicating how often and for what operations the calculator was used each day.

All students were pretested in several important areas. One question under investigation was whether or not manual dexterity enters into eye-hand coordination. Since this can easily affect the student's use of the calculator, MEARC's staff adapted a test to measure dexterity. It was hoped that the study would reveal whether there is a relationship between this factor and the use of the minicalculator.

Since positive student attitude towards mathematics is an important objective of mathematics education, students were tested on a pre/post basis to determine any changes in their attitudes toward mathematics. In addition, the parents of all students in both the experimental and control groups, as well as the mathematics teachers in the West Chester schools, were surveyed concerning their attitudes toward minicalculators in the classroom. Both of these groups were surveyed at the conclusion of the study to determine what, if any, changes had occurred.

As was previously stated, the main question under investigation was the effect of the machine on overall achievement and basic skills. To assess this, all students were pretested on the Cooperative Mathematics Tests (Educational Testing Service). Another form of the test was administered in January 1976, and a third form, which constituted the posttest, in June 1976. The students were *not* permitted to use the calculators during the testing periods.

Finally, to determine if the calculator can aid the student, a second posttest was administered in June 1976 for which the experimental group *was permitted the use of the machine.*

PRELIMINARY FINDINGS

Parent attitude

Table 1 contains selected items from the questionnaire that was mailed to the parents of all seventh-grade students during the summer of 1975. Responses are given in percent form for the 450 questionnaires that were returned, which constituted a 60% response. In addition, parents were given the opportunity to express their opinion at the end of the survey. Although this type of request is often disregarded on surveys, most parents did make comments. Sample comments included—

"It's all right to introduce the calculator in the higher grades, after the students learn their basic skills."

"Let's go back to teaching the basics, not teach our children to be dependent upon a machine."

"Under no circumstances should the taxpayers' money be spent on this."

"The calculators are too easily stolen."

"It could serve as an incentive to the child and add interest to what otherwise might be a dull subject."

"Stop experimenting with our kids; you have already lost one generation to modern math."

"No way our kids should use the machines. Teach them basics."

"It's a good idea! But what will teachers do with the time left over?"

Student achievement

An analysis of the pretest data revealed that the control group was slightly better than the experimental group in both overall achievement and in basic skills. This difference was statistically accounted for when the intermediate test data were interpreted. With the statistical adjustment, a comparison of the groups at the conclusion of the intermediate test indicated that they were not statistically different. The slight

Table 1
Selected items from the parent survey

Question: *In what way might the use of the minicalculator in the seventh-grade classroom affect your child's performance in basic computational skills?*

Hinder: *51%* Have little or no effect: *17%*
Improve: *28%* No opinion: *4%*

Question: *In what way might the use of the minicalculator in the seventh-grade classroom affect your child's overall performance in mathematics?*

Hinder: *34%* Have little or no effect: *22%*
Improve: *39%* No opinion: *4%*

Question: *In what way might the use of the minicalculator in the seventh-grade classroom affect your child's attitude towards mathematics?*

Hinder: *17%* Have little or no effect: *24%*
Improve: *55%* No opinion: *4%*

Question: *Will the use of the minicalculator in the seventh-grade classroom make your child dependent on it in the performance of basic skills?*

Highly dependent: *48%* Dependent to a slight degree: *41%*
Not dependent: *8%* No opinion: *3%*

Question: *Should the West Chester School System permit the use of the minicalculator in the seventh-grade classroom?*

Yes: *49%* No: *43%* No opinion: *8%*

Question: *Should the West Chester School System provide instruction to its students in the use of the minicalculator?*

Yes: *78%* No: *19%* No opinion: *3%*

Question: *If your answer to the previous question was "Yes," in what grade do you think this instruction should begin?*

12th: *5%* 10th: *12%* 8th: *2%* Below 7th grade: *17%*
11th: *7%* 9th: *13%* 7th: *44%*

difference that was revealed favors the experimental group.

CONCLUSIONS

At the time of the writing of this paper, we are able to draw preliminary conclusions in two areas of this study: parent attitude and student achievement.

It is fairly safe to say that the majority of parents participating in the survey had sincere reservations about the introduction of the minicalculator into the classroom. Apparently, the parents felt that the use of the minicalculator in the seventh grade would have a hindering effect on pupil performance in basic skills, but they were not sure how it might affect the overall performance. They felt that the minicalculator might improve pupil attitude towards mathematics, but would have little or no effect on the students' attitude towards school in general. They were overwhelmingly certain that their children would become at least somewhat dependent on the calculators in performing basic skills, but were evenly divided on whether to permit the calculators to be used in the mathematics classroom. It is interesting to note that the overwhelming number who want the use of the calculator to be taught in the schools. Apparently they feel that the students should learn *how* to use the minicalculator as a tool or aid, but should not begin to actually use it until the basic operations have been learned and understood, assuming that basic operations have been taught before the eighth grade.

Although the study is incomplete as this is written, we feel comfortable in saying that the students who have had the use of the calculator *have not at all suffered in either their overall achievement, or their ability to perform paper-and-pencil operations.* Both groups have shown growth, and learning has taken place in both computation and overall mathematical performance. It appears that the experimental group is showing a slight gain in both areas over the control group.

Final data is expected to be available for distribution during the fall of 1976. Interested persons can request data from either of the authors.